普通高等教育“十四五”规划教材

English for Mining Engineering

采矿专业英语

主　编　毛市龙　明　建
副主编　孙金海　李佳洁　韩　斌

北　京
冶　金　工　业　出　版　社
2022

内 容 提 要

本教材共分21章，内容主要包括：地质术语、采矿术语、地下矿山开拓、竖井开拓所具备的条件、主井位置选择、大型金属矿床的地下开拓、金属矿床地下开采、房柱法、全面采矿法、留矿采矿法、分段矿房法、充填法、崩落法、矿井通风，以及露天开采发展历史、露天开采计划与设计、最终开采境界的确定、露天坑境界优化、矿岩运输工艺与排岩工艺、凿岩、爆破等。

本教材为矿业工程专业本科生英语教材，也可供从事矿山设计、研究以及现场生产的工程技术人员和涉外投资采矿的技术人员参考。

图书在版编目(CIP)数据

采矿专业英语 = English for Mining Engineering/毛市龙，明建主编．—北京：冶金工业出版社，2022. 4

普通高等教育“十四五”规划教材

ISBN 978-7-5024-9110-9

Ⅰ．①采… Ⅱ．①毛… ②明… Ⅲ．①矿山开采—英语—高等学校—教材 Ⅳ．①TD8

中国版本图书馆CIP数据核字(2022)第051790号

English for Mining Engineering 采矿专业英语

出版发行	冶金工业出版社	**电　　话**	(010)64027926
地　　址	北京市东城区嵩祝院北巷39号	**邮　　编**	100009
网　　址	www. mip1953. com	**电子信箱**	service@ mip1953. com

责任编辑　郭冬艳　美术编辑　彭子赫　版式设计　郑小利

责任校对　梁江凤　责任印制　禹　蕊

北京虎彩文化传播有限公司印刷

2022年4月第1版，2022年4月第1次印刷

787mm×1092mm　1/16；10. 75印张；259千字；164页

定价 39. 00 元

投稿电话　(010)64027932　投稿信箱　tougao@ cnmip. com. cn

营销中心电话　(010)64044283

冶金工业出版社天猫旗舰店　yjgycbs. tmall. com

(本书如有印装质量问题，本社营销中心负责退换)

前　　言

随着工业革命的全球化进一步推进，越来越多的国内企业走出国门进行矿业投资及采矿活动。为了使国内采矿技术人员更好地适应国际交流，了解国外矿业发展动向，作者根据实际情况，编写了与采矿专业有关的英语教材。该教材紧扣目前国内主流矿床开采工程教材的有关内容，采用英文编写，并针对文章难点编写了英译汉的解释。从而使读者更易于将英文章节中的具体内容与国内汉语教材中的具体内容对照起来，便于理解。本教材作为高校采矿工程专业学生用书，需要读者先学习完地下开采、露天开采、矿井通风、凿岩爆破等相关知识之后，才能理解和掌握本教材相关的英文内容。通过学习采矿工程专业英语的相关章节，读者可以方便地与国外同行进行采矿工程的相关交流。

本教材各章节采用英文介绍地下开采与露天开采的基本知识，大部分章节都配有工程图表，方便读者理解相关的内容。为了方便读者学习，每章结尾都列出了相关的采矿专业名词及解释，同时还对部分重点难点语句进行了英译汉的解释，读者可以在理解重点难点语句后，可掌握其他部分内容的翻译。读者在阅读英文章节时，对照国内矿床地下开采以及露天开采的基本知识，可以较容易地理解采矿专业英语教材的基本内容。

本教材由北京科技大学的毛市龙、明建担任主编，孙金海、李佳洁、韩斌担任副主编，高永涛、明世祥、张磊担任主审工作。在本教材的编写过程中得到了中南大学的何哲祥、武汉科技大学的钟冬望、北京理工大学的杨军细心指点与帮助，感谢他们的辛勤劳动。

在编写过程中，参阅了国内外采矿工程专业方面的有关资料，在此向原作者和相关人员表示衷心的感谢。

本教材的编写和出版，得到了教育部本科教学工程专业综合改革试点项目经费和北京科技大学教材建设基金的资助。

由于编者水平所限，不足之处，诚请广大读者批评指正。

编　者
2022 年 1 月

前言

[illegible]

[illegible]

[illegible]

[illegible]

[illegible]

[illegible]

[illegible]

目　　录

Unit 1 Geological Terminology

地质术语

1. Introduction

There are some general geological terms that should be understood for exploration and mining purposes. The simplest form of rock strata is that in which each stratum (layer) of rock is superimposed on another. These are conformable beds. Movements of the earth's crust and subsequent weathering and rock displacement provide conditions for mineral formation and relocation into the site where it will be found.

The original layers of rock were formed by sedimentary or chemical processes with material from higher land masses being redeposited in lower areas. The beds could then be distorted into folds to produce synclines or anticlines. The scale of those movements is such that a single fold producing a synclinal basin may stretch for hundreds of kilometers or may be measured with a wavelength of a few meters.

Where the folding is severe the rock is frequently heavily fractured, particularly where it is under tension on the crest of an anticlinal stratum or at the base of a synclinal stratum. These weakened zones provide a path for igneous intrusive rocks or mineral rich solutions to permeate the host rock and produce ore bodies. It is for this reason that many ore bodies have a typical form, such as the saddle reefs. In practice a fault of this magnitude would have many associated smaller faults and fractures that can again give rise to an area favourable for subsequent mineralisation.

An unconformity is also shown in this case produced by an old shore line. These shore lines were often the location of the deposition of coal seams in the younger geological eras. It can be seen that it is important to distinguish where a coal seam has terminated at the edge of a basin or whether it has been interrupted by a fault or lost by erosion at an outcrop. Outcrops occur when a series of beds are cut away by general erosion of the surface or dissected by rivers, so that the edges of beds are exposed.

After strata have been deposited and mildly disturbed they may be broken up orogenic movements. These are the large movements of the earth's crust that are associated with mountain building and sometimes cause eruption or intrusion of igneous rock with subsequent mineralisation.

Syngenetic deposits are those mineral deposits formed contemporaneously with the parent rock and enclosed by it. They can be igneous or sedimentary. Epigenetic deposits are formed subsequently to the enclosing rock, often by a solution (the mother liquor) penetrating into rocks and depositing minerals. If the solution is a result of magmatic (igneous) action it will probably be hot and under pressure and will deposit non-ferrous minerals. If the solution is a result of sedimentary action it will percolate downwards causing either chemical changes or direct deposition. Lateritic deposits are formed by the leaching away of soluble minerals leaving behind in the laterite a valuable ore, such as nickel or bauxite (aluminium ore). Thus these are normally surface deposits. Although some, such as the iron ores, have been changed and buried so that they are no longer lateritic.

2. Geological Terminology

(1) Ore and Gangue. An orebody may contain both desirable metallic minerals and unwanted non-metallic minerals. The metallic minerals form the ore and the unwanted minerals are known as gangue.

(2) Mullock. An orebody should be known more accurately as mineralisation which comprises economic ore and sub-economic lower grade material. The lower grade material broken during mine development is called mullock or waste. These terms are sometimes extended to include host rock broken in development drives.

(3) Disseminated and Massive ores. Where the ore occurs in relatively pure form for example as almost 100 percent lead sulphide (galena) in even small amounts it is said to be massive. If however the ore is spread thinly through a body of rock so that individual specks of the ore are surrounded by a high percentage of rock then this is a disseminated ore. Thus if an ore body is sampled over a narrow width of massive sulphides it could show a value of lead in the ore sample of 50 to 60 percent (lead in pure galena is in fact 86.6 percent). Over a greater width it might only show about 20 percent lead because the sulphides had become disseminated or dispersed. The disseminated sulphides, assayed separately (i. e. analysed for metal content), may only show as little as 0.2 to 0.5 percent lead.

(4) Cut-off-grade. A gradual transition of high grade to low grade ore is common and it is not always easy to determine the physical limits of an ore body. Consequently it is necessary to determine the lowest value of metallic or non-metallic mineral content which can be mined economically and use this as the physical limit of the orebody. However this limit will obviously vary with the market value of the ore and the mining methods.

(5) Gossan. Where orebodies outcrop to the surface, particularly in geologically mature areas free of glacial deposition, the portion near to the surface may be leached by impure percolating waters and subsequently oxidized. Many metallic ores contain a proportion of iron sulphides and these do not leach away but change to iron oxides (haematite, goethite and jarosite). The colour of the oxides varies slightly with the original mineral content, but general end result is a yellowish, brownish, or reddish ferruginous capping known as a gossan. This is generally situated directly over the ore body in arid area, such as in Australia, but in some cases it is transported from the outcrop and has to be traced back.

(6) Supergene Enrichment. The minerals leached away from the outcrop often cause a zone of

secondary enrichment (generally in the form of sulphides) at a depth of 60 to 120m from the surface. This enrichment zone then lies on top of the unaltered primary orebody. Consequently if one in mining or prospecting from the surface for non-ferrous metallic orebodies, there is consequently a transition from oxidized ores, which may need special extractive metallurgical processes, to the supergene enrichment zone showing high metal contents, and then to the orebody proper which may have a lower metal content again. Thus prospecting with shallow drill holes can give very misleading results.

(7) Placer and Alluvial Deposits. When the primary deposits are brought to the surface by geological methods and then mechanically eroded, they can suffer one of two fates. They may go into solution for subsequent chemical redeposition or they may break up into fairly stable particles which are transported varying distances. Heavy mineral particles, such as gold, settle close to their source. The original falling area, known sometimes as a rock scree when situated against a mountain slope, is the eluvial region. Further transportation produces the placer deposits of such as gold and tin, although there appears to be no precise separation of this term from that of alluvial deposits. The alluvial deposits may be found in existing river beds or in old river beds covered with new deposits such as basalt flows. These buried deposits may be called deep leads as in the state of Victoria, Australia, where much of the gold of the 1850's rush was found.

Transportation to the sea followed by wave action produces beach deposits. Mineral sands is the term applied to the heavy sand deposits of rutile, ilmenite and zircon, although they are sometimes colloquially called beach sands. These can be on either existing or old shore lines. Other minerals such as tin or diamonds also are found in beach deposits.

(8) Bedded Deposits. Chemical and physical deposition of consolidated minerals can produce bedded deposits which may extend over larger areas. These are minerals such as coal, iron ore, salt or potash deposits, and phosphates. Limestone required for cement or blast furnace is also a bedded mineral deposit. Although the dip (slopes) of these bedded deposits can be steep, and the beds are broken by faults or other geological disturbances, their lateral extent is usually larger and more continuous than is that of the metallic non-ferrous deposits.

Vocabulary

superimposed [ˌsjuːpərimˈpəuzd] *adj.* 叠置的；叠积的
sedimentary [ˌsediˈmentəri] [ˌsediˈmentri] *adj.* 沉淀物的
syncline [ˈsɪŋklaɪn] *n.* 向斜
fold [fəuld] *vt.* 折叠；
n. 折层；折页；折痕
crest [krest] *n.* 鸟冠；顶
vt. 到达……之顶
vi. 形成顶；到达顶端
igneous [ˈigniəs] *adj.* 火成的

permeate [ˈpəːmieit]	v. 透过；渗透
unconformity [ˌʌnkənˈfɔːmiti]	n. 不一致；不相合 [地理] 地层不整合
seam [siːm]	n. 缝，接缝；层
	vt. 缝合；接合
	vi. 产生裂口
deposition [ˌdepəˈziʃən]	n. 罢免；储存；沉积物
fault [fɔːlt]	n. 缺点；[地理] 断层
outcrop [ˈautkrɔp]	vi. 露出地面；出现
	n. （矿脉）露出地面的部分；露头
deposit [diˈpɔzit]	n. 沉淀物；沉淀
epigenetic [epɪdʒɪˈnetɪk]	adj. 外力变质的
magmatic [ˈmægmətɪk]	adj. 岩浆的；乳剂的
non-ferrous [ˈnʌnˈferəs]	adj. 有色的；非铁或钢的（金属）
percolate [ˈpəːkəleit]	vi. 渗透
	vt. 渗透过
lateritic [lætəˈritik]	adj. 含红土的；有红土之特性的
gangue [gæŋ]	n. 脉石（=gang）；矿皮；弃矿
mullock [ˈmʌlək]	n. 矿山废石
galena [gəˈliːnə]	n. 方铅矿
sulphide [ˈsʌlfaid]	n. [化学] 硫化物
speck [spek]	n. 污点
	vt. 使玷污点；使产生斑点
massive [ˈmæsiv]	adj. 巨大的；[矿物] 非结晶形的
dissect [diˈsekt]	vt. 解剖；详细研究
	vi. 解剖
orogenic [ˌɔrəuˈdʒenik]	adj. 造山运动的，造山作用的
gossan [ˈgɔsn]	n. [矿] 铁帽
haematite [ˈhemətait]	n. =hematite，[矿物] 赤铁矿
goethite [ˈgəuθait]	n. 针铁矿
jarosite [ˈdʒærəsait]	n. [矿] 黄钾铁矾
ferruginous [feˈruːdʒinəs]	adj. 铁的；含铁的；铁锈色的；赤褐色的
supergene [ˈsjuːpədʒiːn]	adj. 浅成生的，表生的
	n. 浅生矿床
transition [trænˈziʒən]	n. 转变；过渡；[地理] 转移层
placer [ˈpleɪsə]	n. 砂矿
alluvial [əˈluːviəl]	n. 冲积土
	adj. 冲积的；冲积期的
solution [səˈluːʃən]	n. 溶解；分离
redeposition [ˈrɪːdepəˈzɪʃən]	n. 再沉积

scree [skriː] n. 山麓碎石；卵石；岩屑堆
eluvial [ɪ'ljuːvɪəl] adj. 残积层的
basalt ['bæsɔːlt] n. 玄武岩；一种黑色的陶器
rutile ['ruːtil] n. 金红石
ilmenite ['ilmənait] n. 钛铁矿
zircon ['zəːkɔn] n. 锆石（用作宝石）
potash ['pɔtæʃ] n. 钾碱；碳酸钾；钾；氢氧化钾
phosphate ['fɔsfeit] n. 磷酸盐；磷酸；磷肥
conformable beds 整合层
anticlinal stratum 背斜地层
saddle reef 鞍礁
syngenetic deposit 同生矿床
epigenetic deposit 后生矿床
sedimentary action 沉积作用
disseminated ore 浸染状矿石
parent rock 母岩
broke up 分手
glacial deposition 冰川沉积
supergene enrichment 表生富集
non-ferrous metallic 有色金属
placer and alluvial deposits 砂矿和冲积矿床
bedded deposit 层状矿床

NOTES

[1] There are some general geological terms that should be understood for exploration and mining purposes. The simplest form of rock strata is that in which each stratum (layer) of rock is superimposed on another. These are conformable beds. Movements of the earth's crust and subsequent weathering and rock displacement provide for mineral formation and relocation into the site where it will be found.

了解一些地质术语有利于学习地质勘探和采矿知识。地下岩层最简单的形式是一层岩石叠加在另一层岩石上。这些岩层属整合地层。矿物是随着地壳的运动以及随后的风化和岩石移动而形成的，矿物随后迁移并聚集到当前被发现的地点。

[2] The original layers o rockwere formed by sedimentary or chemical processes with material from higher land masses being redeposited in lower areas. The beds could then be distorted into folds to produce synclines or anticlines. The scale of those movements is such that a single fold producing a synclinal basin may stretch for hundreds of kilometers or may be measured with a wavelength of a few meters.

高处的岩体滚落到低处，通过沉积作用和化学作用最终形成了原始岩层。之后，这些沉积岩层可能会被扭曲成褶皱，从而产生向斜或背斜。这些岩层扭曲运动的规模有大有小，大的向斜盆地的单个褶皱可能延伸数百公里，而小的只有几米。

[3] Where the folding is severe the rock is frequently heavily fractured, particularly where it is under tension on the crest of an anticlinal stratum or at the base of a synclinal stratum. These weakened zones provide a path for igneous intrusive rocks or mineral rich solutions to permeate the host rock and produce ore bodies. It is for this reason that many ore bodies have a typical form, such as the saddle reefs. In practice a fault of this magnitude would have many associated smaller faults and fractures that can again give rise to an area favourable for subsequent mineralisation.

在褶皱严重的地层，岩石经常出现严重的节理裂隙，特别是在背斜地层顶部或向斜地层底部受到拉伸的岩层其裂隙更加发育。这些节理裂隙发育的弱化带为火成岩或含矿熔岩侵入母岩体，并为成矿提供了通道。正是由于这个原因，许多矿体都有一个典型的形态，如鞍礁状形态。实际上，这种规模的断裂带都伴随有许多较小的断层和裂隙，之后，这些小的断层和裂缝又为成矿提供了条件。

[4] An orebody should be known more accurately as mineralisation which comprises economic ore and sub – economic lower grade material. The lower grade material broken during mine development is called mullock or waste. These terms are sometimes extended to include host rock broken in development drives.

矿体更准确地说就是矿化体，它包括了经济矿石和低品位次经济矿石。矿山开采过程中产生的低品位次经济矿石也称为含矿废石或废石。在采矿过程中，在围岩里掘进时产生的岩石也叫废石。

[5] Where the ore occurs in relatively pure form for example as almost 100 percent lead sulphide (galena) in even small amounts it is said to be massive. If however the ore is spread thinly through a body of rock so that individual specks of the ore are surrounded by a high percentage of rock then this is a disseminated ore. Thus if an ore body is sampled over a narrow width of massive sulphides it could show a value of lead in the ore sample of 50 to 60 percent (lead in pure galena is in fact 86.6 percent). Over a greater width it might only show about 20 percent lead because the sulphides had become disseminated or dispersed. The disseminated sulphides, assayed separately (i. e. analysed for metal content), may only show as little as 0.2 to 0.5 percent lead.

如果矿石以相对纯的形式出现，即使是少量的近乎 100% 的方铅矿，也称为块矿。当矿石稀疏地散布在一块岩石上，其中的矿石斑点被大量岩石包围，这就是一种浸染状矿石。因此，如果在块状硫化物矿石的局部进行取样，矿石样品中的铅含量可达 50% ~60%(实际上纯方铅矿中的铅含量为 86.6%)。而在更宽的范围内取样化验，则矿体的铅含量可能只有 20%，这是因为硫化物矿石已经扩散或分散在岩石中去了。而在浸染状矿体里取样化验（即按金属含量分析)，其含铅量可能只有 0.2%~0.5%。

[6] A gradual transition of high grade to low grade ore is common and it is not always easy to determine the physical limits of an ore body. Consequently it is necessary to determine the lowest value of metallic or non-metallic mineral content which can be mined economically and use this as the physical limit of the orebody. However this limit will obviously vary with the market value of the ore and the mining methods.

通常在矿体中，矿石品位从高到低逐渐过渡，确定矿体的界限就比较困难。因此，有必要确定经济上合理的最低的矿石品位，并以此作为矿岩的区分界线。当然，这一界线会随矿石的市场价值和采矿方法而发生改变。

Unit 2 Mining Terminology

采矿术语

1. Large Deep Orebodies

Large ore-bodies are often known as massive, but the term massive should not be confused with massive and disseminated ores in a geological sense. In fact a massive ore-body is likely to contain disseminated ore minerals. A vein-type ore-body will probably contain bands of massive ore minerals.

A shaft is sunk alongside or through the ore-body for access. In practice, there must be two openings to allow air to be forced through the mine workings. Provided that there are at least two means of access to the mine further shafts may be partitioned for two-way airflow within them, but this is not permitted in coal mines. It must be noted that if the shaft is sunk through the ore-body then some of the ore is sterilized. A cone of ore surrounding the shaft and widening with depth must be left until the rest of the ore is extracted. Otherwise the movement of rock caused by ore extraction will damage the shaft and necessitate expensive repairs, and lost production.

Waste rock, ore, men and materials have to be raised and lowered in the shafts. There are two types of conveyance, cages and skips. Cages are used for man-riding and to wind mine cars containing ore, waste rock or materials.

A large shaft may be equipped with a pair of skips for ore and rock hoisting and a single cage for man riding. The area around the shaft at ground level is known as the brace in metal mining and pit top or bank in coal mining.

At regular intervals down the shaft, openings are made and levels are driven into the orebody to enable it to be worked. The openings are called plats in Australia, stations in South Africa and the U. S. A, and insets in Britain. The interval between the levels is a function of the mining method and may vary from about 25 to 60m. At Mount Isa for instance the level interval is approximately 55m with every second level being a haulage level. Improved drilling and raising techniques have enabled the interval between main levels to be increased in newer developments in mines.

Within the ore-body it may be necessary to have more frequent access for working purposes, and sublevels will be driven at intervals between the main levels. It is also necessary to make vertical connections between levels for ventilation access for men and materials (known as service raises), and for dropping ore or rock through (ore passes). The connections may be made by driving upwards, known as winzing. These developments are made in a vertical plane and must be connected to each other in a horizontal plane by crosscuts, or drives.

After the ore-body has had some or all of its development openings made, partly in barren rock and partly in the ore-body, then systematic extraction of the profitable mineral starts. The working area in which mining is taking place is known as stope.

2. Vein-type Ore-body

A large ore-body may in fact consist of clusters of smaller ore-bodies of a vein-type, and development may be on a regular grid. Alternatively the ore-body may be one predominant vein which is long and narrow on its strike (horizontal) direction and which extends to the dip for a considerable depth. Technically the strike is at right angles to the full dip, but within an orebody the strike may follow a sinuous line and the dip frequently varies so that general trends have to be taken.

It can be seen that a vertical shaft rapidly loses contact with an inclined orebody and increasingly long levels have to be driven. It can also be appreciated that mining operations would weaken the hanging wall i. e. the rock lying above the ore body, and cause it to collapse. As a result the major developments are normally made in the footwall, below the orebody, where the rock is more stable.

In the early stages of working the shallowest ore an inclined shaft may be driven down from the surface and progressively deepened to follow ore extraction. However haulage speeds are generally lower and maintenance costs are higher in inclined shafts, so that with increased depth a change will be made to vertical shaft winding. An ore pass system will probably be developed to drop ore and rock to widely spaced main haulage levels.

In areas such as the Coeur Alene silver / lead district in the U. S. A. and gold mining areas in the Witwatersnand, South Africa, where workings in veins or reefs have extended to 2500 or 4000 meters, it has been found that single shaft hoisting is uneconomic. Consequently the shafts are broken into lengths of about 1000 meters to 1500 meters and the surface winding gear is duplicated underground. Even with modern technology this is still necessary or may be more convenient for inclined orebodies where shafts are progressively deepened. For near-horizontal bedded deposits on a single horizon, where the full depth has to be sunk before production of mineral can start, then single lifts of over 2000 meters are made.

3. Large Shallow Orebodies

Shallow orebodies are worked as open pit mining operations wherever possible. The scale of equipment is generally smaller in metalliferous orebodies than in shallow bedded deposits because of the need for benching downwards (To start the removal of each successive layer a center cut, or drop cut, is made in the pit bottom and a bench is taken outward).

Because of the need to keep the pit slopes stable the overall inclination of the pit walls is generally about 45° and bench heights are generally about 15m. it can be seen that as the pit becomes deeper then more waste rock (overburden, or spoil) has to be taken in ratio to the ore won. Eventually economics force a change to underground mining methods.

Haulage out of the pit is at first by truck in a spiral around the benches. It may then become more economic and convenient to drop the ore through a crusher to an underground haulage for transport out of the mine. The access tunnel driven down through the strata to the bottom of the open is called a drift. A drift is usually driven as a straight line in plan to reach from the surface to the desired underground horizon and is used for rail or conveyor haulage at grades of up to 1 in 2. 5. With locomotive haulage the grade are restricted to less than 1 in 15.

With a change to underground methods of mining, two types of haulage are possible. Down to about 200 meters it is possible to haul ore economically by low-profile diesel trucks up an access spiral. This access is driven at a grade of about 1 in 7 to 1 in 9 and in addition ore haulage allows diesel driven rubber-typed loaders and ancillary equipment to be taken into the mine and brought out for servicing. For large scale mechanized mining this is probably cheaper than the alternative of hauling ore and waste rock to a shaft and then hoisting it to the surface. However below 200 to 250m a detailed study will probably show that shaft hoisting is cheaper than trucking up an incline, although the spiral incline will be continued for vehicle access and through ventilation.

Vocabulary

vein [vein]	*n.* 矿脉;
	vt. 在……画条纹；提供脉络
shaft [ʃaːft]	*n.* 井筒
sink [siŋk]	*vi.* 下沉
	vt. 使下沉；钻
partition [paːˈtiʃən]	*n.* 划分；隔墙
	vt. 分割；隔开
sterilize [ˈsterilaiz]	*vt.* 使不能生产；消毒
cage [keidʒ]	*n.* 罐笼，监狱
	vt. 关入笼中；监禁
skip [skɪp]	*n.* 箕斗
brace [breis]	*n.* 支柱；紧缚之物
	vt. 支住
bank [bæŋk]	*n.* 岸
	vt. 围以堤
level [ˈlevl]	*n.* 中段，[矿物] 主平巷
	adj. 水平的
plat [plæt]	*n.* 褶皱；辫状物（=plait）
inset [ˈinˈset]	*vt.* 嵌入；插入
sublevel [ˈsʌblevəl]	*n.* [矿物] 分段
winzing [winzɪŋ]	*n.*（垂直或倾斜的）开暗井
crosscut [ˈkrɔskʌt]	*v.* 横越；横切

	n. 切割横巷
stope [stəup]	n. 采场
cluster ['klʌstə]	n. 串；束；丛；群；团
alternatively [ɔːl'təːnətivli]	adv. 二选一地，替换地，选择地
predominant [pri'dɔminənt]	adj. 主要的；有势力的
strike [straik]	n. 走向
sinuous ['sinjuəs]	adj. 弯弯曲曲的；蜿蜒的
appreciate [ə'priːʃieit]	vt. 为……表示感激；重视
	vi. 涨价；增多
footwall ['fʊtwɔːl]	n. 底板
extraction [iks'trækʃən]	n. 抽出；采矿
progressively [prə'gresivli]	ad. 日益增多的
metalliferous [ˌmetə'lifərəs]	adj. 产金属的；含金属的
bench [bentʃ]	n. 台阶
overburden [ˌəuvə'bəːdn]	vt. 使负担过重
	n. 过重的负担［矿物］上覆岩层
bands of	成群的
barren rock	贫瘠岩石，围岩
hanging wall	顶板
inclined shaft	斜井
ore pass	矿石溜井
winding gear	卷绕装置
low-profile diesel truck	高度矮的柴油卡车
rubber-typed loader	轮式装载机
ancillary equipment	辅助设备
vein-type ore-body	脉状矿体
disseminated ores	浸染状矿石

NOTES

[1] Large ore-bodies are often known as massive, but the term massive should not be confused with massive and disseminated ores in a geological sense. In fact a massive ore-body is likely to contain disseminated ore minerals. A vein-type ore-body will probably contain bands of massive ore minerals.

大型矿体通常为块状矿体，但是这里的块状矿体与地质学中的块状矿石与浸染状矿石不是一个概念。实际上，块状矿床通常包含浸染状的矿石。脉状矿体可能包含若干条带状的浸染状矿石。

[2] A shaft is sunk alongside or through the ore-body for access. In practice there must be two

openings to allow air to be forced through the mine workings. It must be noted that if the shaft is sunk through the ore-body then some of the ore is sterilized. A cone of ore surrounding the shaft and widening with depth must be left until the rest of the ore is extracted. Otherwise the movement of rock caused by ore extraction will damage the shaft and necessitate expensive repairs, and lost production.

要进入矿体开采，首先要在矿体周边掘进竖井或在矿体中掘进竖井作为出入口。实际生产过程中，矿山必须具备两个出入口，以便让空气通过井下工作面。需要注意的是，当竖井穿过矿体时，则需要保留部分矿体作为保安矿柱。在矿体其他部分开采完之前，竖井周边这一下大上小的锥形保安矿柱必须保留。否则，矿石开采引起的岩石移动将破坏竖井，需要进行的维修费昂贵，并且影响产量。

[3] Waste rock, ore, men and materials have to be raised and lowered in the shafts. There are two types of conveyance, cages and skips. Cages are used for man-riding and to wind mine cars containing ore, waste rock or materials.

废石、矿石、人员和材料都必须经过竖井提升。竖井提升有两种方式：罐笼提升和箕斗提升。罐笼提升人员、矿石车和废石车以及材料。

[4] A large shaft may be equipped with a pair of skips for ore and rock hoisting and a single cage for man riding. The area around the shaft at ground level is known as the brace in metal mining.

大竖井可以安装一对箕斗用以提升矿石和岩石，还可以安装一个罐笼提升人员。在金属矿山，井筒地面部分称井架。

[5] A large ore-body may in fact consist of clusters of smaller ore-bodies of a vein-type, and development may be on a regular grid. Alternatively the ore-body may be one predominant vein which is long and narrow on its strike (horizontal) direction and which extends to the dip for a considerable depth. Technically the strike is at right angles to the full dip, but within an orebody the strike may follow a sinuous line and the dip frequently varies so that general trends have to be taken.

通常，一个大矿体可以是许多小的矿脉组成，因此，矿山开拓需要全盘考虑。矿体也可能是一个走向长，宽度窄，向下延伸深的主矿脉。技术上说，矿体走向与倾向垂直。但是矿体的走向往往不是一条直线，而是弯曲向前，倾向也常常发生改变，因此，在设计中常用采用矿体的大致走向和倾向来说明矿体的赋存情况。

[6] It can be seen that a vertical shaft rapidly loses contact with an inclined orebody and increasingly long levels have to be driven. It can also be appreciated that mining operations would weaken the hanging wall i. e. the rock lying above the ore body, and cause it to collapse. As a result the major developments are normally made in the footwall, below the orebody, where the rock is more stable.

通常，竖井越深，它与矿体的距离就越远，从竖井开掘石门至矿体的长度就越长。同样，矿床开采将削弱上盘围岩（即矿体上部的岩层），导致上盘围岩发生破坏。因此，主要的开拓工程往往布置在矿体下盘，因为下盘围岩更加稳定。

[7] In the early stages of working the shallowest ore an inclined shaft may be driven down from the surface and progressively deepened to follow ore extraction. However haulage speeds are generally lower and maintenance costs are higher in inclined shafts, so that with increased depth a change will be made to vertical shaft winding. An ore pass system will probably be developed to drop ore and rock to widely spaced main haulage levels.

在早期开采阶段，埋深最浅的矿体可以从地面掘井斜井。随着矿石的开采，逐渐延伸斜井。但是，斜井开拓时的提升速度比较低，维修成本高，因此，随着开采深度的增加，提升方式就需要进行调整，由斜井提升变为竖井提升。如果开拓系统中增设溜井出矿系统用以出矿和出岩，那么中段高度可以提高。

Unit 3 Underground Mine Development

地下矿山开拓

1. Introduction of Mine Development

Types of underground development openings can be ranked in three categories by order of importance in the overall layout of the mine: (1) Primary: main openings (e. g., shaft, slope); (2) Secondary: level or zone openings (e. g., drift, entry); (3) Tertiary: lateral or panel openings (e. g., ramp, crosscut).

Generally, the openings are driven in this order; that is, from main development openings to secondary level or zone openings, to tertiary openings like lateral or panel openings. However, many variations exist with different mining methods. For example, in coal mining, the entries and the associated crosscuts are always driven at the same time, regardless of the category of the entries. It should also be noted that underground mining often employs a distinctive nomenclature. The following lists define a number of the terms commonly used to describe underground workings and other aspects of underground mining.

(1) Deposit and spatial terms

1) back: roof, top, or overlying surface of an underground excavation.

2) bottom: floor of underlying surface of an underground excavation.

3) capping: waste material adjacent to a mineral deposit.

4) crown pillar: portion of the deposit overlying an excavation and left in place as a pillar.

5) dip: angle of inclination of a deposit, measured from the horizontal.

6) floor: bottom or underlying surface of an underground excavation.

7) footwall: wall rock under the deposit.

8) gob: broken, caved, and mined-out portion of the deposit.

9) hanging wall: wall rock above a deposit.

10) pillar: unmined portion of the deposit, providing support to the roof or hanging wall.

11) rib: side wall of an excavation; also rib pillar.

12) roof: back, top, or overlying surface of an excavation.

13) sill pillar: portion of the deposit underlying an excavation and left in place as a pillar.

14) strike: horizontal bearing of a tabular deposit at its surface intersection.

15) overhand: advancing in an upward direction.

16) underhand: advancing in a downward direction.

(2) Excavation terms

1) bell: funnel-shaped excavation formed at the top of a raise to move bulk material by gravity from a stope to a drawpoint.

2) bleeder: exhaust ventilation lateral.

3) chute: opening from a drawpoint, utilizing gravity flow to direct bulk material from a bell or orepass to load a conveyance.

4) crosscut: tertiary horizontal openings, often connecting drifts, entries, or rooms; oriented perpendicularly to the strike of a pitching deposit; also breakthrough.

5) decline: secondary inclined opening, driven downward to connect levels, sometimes on the top of the dip of a deposit; also declined shaft.

6) drawpoint: loading point beneath a stope, utilizing gravity to move bulk material downward and into a conveyance, by a chute or loading machine, also boxhole.

7) drift: primary or secondary horizontal or near-horizontal opening, oriented parallel to the strike of a pitching deposit.

8) finger raise: vertical or near-vertical opening used to transfer bulk material from a stope to a drawpoint, often an interconnected set of raises.

9) grizzly: coarse screening or scalping device that prevents oversized bulk material from entering a material transfer system; constructed of rails, bars, beams, etc.

10) haulage way: horizontal opening used primarily for materials handling.

11) incline: secondary inclined opening, driven upward to connect levels, also inclined shaft.

12) lateral: secondary or tertiary horizontal opening, often parallel or at an angle to a haulage way, usually to provide ventilation or some auxiliary service.

13) loading pocket: transfer point at a shaft where bulk material is loaded by bin, hopper, and chute into a skip.

14) longwall: horizontal exploitation opening several hundred feet in length, usually in a tabular deposit.

15) manway: compartment of a raise or a vertical or near-vertical opening intended for personnel travel between two levels.

16) orepass: vertical or near-vertical opening through which bulk material flows by gravity.

17) raise: secondary or tertiary vertical or near-vertical opening, driven upward from one level to another.

18) ramp: secondary or tertiary inclined opening, driven to connect levels, usually in a downward direction, and used for haulage.

19) room: horizontal exploitation opening, usually in a bedded deposit.

20) shaft: primary vertical of near-vertical opening, connecting the surface with underground workings, also vertical shaft.

21) slope: primary inclined opening, usually a shaft, connecting the surface with underground workings.

22) slot: narrow vertical or inclined opening opening excavated in a deposit at the end or a stope to provide a bench face.

23) stope: large exploitation opening, usually inclined or vertical, but may also be horizontal.

24) sublevel: secondary or intermediate level between main levels or horizons, usually close to the exploitation area.

25) transfer point: location in the materials-handling system, either haulage or hoisting, where bulk material is transferred between conveyances.

26) tunnel: main horizontal or near-horizontal opening, with access to the surface at both ends.

27) undercut: low horizontal opening excavated under a portion of a deposit, usually a stope, to induce breakage and caving of the deposit; also a narrow cut in the face of a mineral deposit to facilitate breakage.

2. Mine Development and Design

Mine development must proceed considering all aspects of the overall mine design. Because of the complexity and expense of underground mining, extreme care must be exercised in making decisions during development that may also affect subsequent production operations. The most crucial matters of concern are discussed in the following sections.

(1) Mining method. Once the decision has been made to sue underground mining, attention is focused on the selection of an exploitation method. Development should not proceed until a mine production plan has been adopted, and the first step is to decide which class of underground method is most suitable: unsupported, supported or caving. Note also that solution mining, although not common, is still a choice for an underground deposit. Selection of a mining method hinges on natural and geologic conditions related to the mineral deposit, on certain economic and environmental factors, and on other factors that may have a bearing on the specific deposit to be exploited.

The reason that the choice of a method is so crucial is that it largely governs the type and placement of the primary development openings. If disturbance of the surface due to subsidence is anticipated—inevitable with caving methods and possible with other methods—then all the access openings must be located outside the zone of fracture bounded by the angle of draw. If the integrity of the ground overlying the active mining area can be ensured for the life of the mine, then the primary openings can be located more centrally above the deposit.

(2) Production rate and mine life. A variety of geologic and economic conditions determine the optimum rate of production from a miner deposit of known reserves and hence, the life of the mine. These include market conditions and selling price of the commodity, mineral grade, development time, mining costs, means of financing, government support and taxation policies, and a number of other factors.

The most common goal in modern mine planning is to optimize the net present value, that is, to mine the deposit so that the maximum internal rate of return is achieved. All other things being equal, the higher the production rate, the shorter the mine life.

3. Main Access Openings

A number of initial decisions related to the primary development openings if a mine must be made early in the mine planning stage. They concern the type, number, shape, and size of the main openings. By necessity, these decisions are ordinarily made at the time when the primary materials-handling system is chosen. Factors influencing the decisions include the depth, shape, and size of the deposit; surface topography; natural and geologic conditions of the orebody and surrounding rock; mining method; and production rate. Wise decision made at the outset avoids later changes in development opening, changes that are always disruptive and expensive. Such changes are common in mines with great longevity but should not occur in short-lived operations.

4. Types of Openings

In regard to types of openings, there are only three common choices that are used with regularity: (1) shafts (vertical and near-vertical); (2) declines (slopes or ramps); (3) adits or drifts. Vertical shafts have always been among the most common types of openings for deep mines. However, nearly all shafts are now vertical because near-vertical shafts are more costly and difficult to develop.

Declines have been among the fastest growing primary development openings because of their association with principal materials-handing systems. In coal mines, these openings are called slopes and are associated with the belt conveyor transport of coal out of the mine. Except in very deep mines, these openings provide for low-cost primary transportation of coal out of mines, eliminating the need for a hoisting system. The slopes in a coal mine are generally driven at an angle of about 15°. For this reason, a slope is usually 3.6 times as long as a vertical shaft and costs more than a shaft. However, a slope will pay for the extra cost in a few years with the savings to be obtained by substituting a belt conveyor for a hoist. Similar logic can be applied to a ramp in a metal mine. These openings are ordinarily driven to allow free access to any level of mine with diesel-powered equipment. They can be driven in a spiral form or in a rectangular form, with linear sections alternating with curved sections to achieve the desired access to each level of the mine. In each case the idea is to provide a means of utilizing mobile equipment throughout the mine without limitation. Ramps are not ordinarily used for primary haulageways. But rather as necessary development openings to facilitate use of diesel equipment in the mine.

Drifts or adits are used in coal, nonmetal, and industrial minerals operations in any situation where the deposit can be easily accessed through entries or adits of a horizontal type. The proper situation would normally occur in coal or nonmetal deposits, but can be readily found in metal deposits located in mountainous regions. Clearly, this method of development can apply to many deposits and is often a cost-effective method of developing a mine if the possibility exists.

Vocabulary

nomenclature [ˈnoumənˌkletʃə]　　n. 专门名词
adit [ˈædɪt]　　n. 入口，通路 [矿] 平巷，矿坑石门
drawpoint [dˈrɔːpɔɪnt]　　n. 出矿口，出矿点
bleeder [ˈblidə]　　n. 易出血的人，回风巷道
chute [ʃut]　　n. 斜槽
　　vt. 以斜槽运送
　　vi. 以斜槽下滑
shaft [ʃaːft]　　n. 井筒；井
stope [stəʊp]　　n. 采场
decline [dɪˈklaɪn]　　n. 下降；斜坡
slope [sləʊp]　　n. (山的) 斜坡；斜面；斜度；斜率；斜井
　　vi. 有坡度
　　vt. 使倾斜
ramp [ræmp]　　n. 匝道；斜坡道
mine development　　矿床开拓
primary haulageway　　主运输巷道
mine life　　矿山开采期
unsupported method　　空场法
supported method　　充填法
caving method　　崩落法

NOTES

[1] Types of underground development openings can be ranked in three categories by order of importance in the overall layout of the mine: (1) Primary: mainopenings (e.g., shaft, slope); (2) Secondary: level or zone openings (e.g., drift, entry); (3) Tertiary: lateral or panel openings (e.g., ramp, crosscut).

地下开采的开拓工程根据其重要程度可分为三类：(1) 主要开拓工程 (如：竖井，斜井)；(2) 水平中段巷道或采场巷道 (如：平巷，进路)；(3) 辅助巷道或盘区巷道 (如：斜坡道，切割巷道)。

[2] Generally, the openings are driven in this order; that is, from main development openings to secondary level or zone openings, to tertiary openings like lateral or panel openings. However, many variations exist with different mining methods. For example, in coal mining, the entries and the associated crosscuts are always driven at the same time, regardless of the category of the entries. It should also be noted that underground mining often employs a distinctive nomenclature.

The following lists define a number of the terms commonly used to describe underground workings and other aspects of underground mining.

通常地下工程是从主要开拓工程开始向水平中段巷道或采场巷道方向进行掘进，再向辅助巷道或盘区巷道进行掘进。但是，不同的采矿方法其掘进的先后次序也发生变化。例如，在煤炭开采中，进路与相应的切割横巷常常是同时掘进的，并没有区分先后次序。我们在学习中还要注意地下开采的特殊的名词术语。下面列出了一些普通地下采矿术语的定义解释。

[3] Deposit and spatial terms 矿床和矿床开采的术语：

1) back：roof, top, or overlying surface of an underground excavation.
顶板：地下采场空间的顶部。

2) bottom：floor of underlying surface of an underground excavation.
底板：地下采场空间的底部。

3) capping：waste material adjacent to a mineral deposit.
围岩：矿床周围的岩石。

4) crown pillar：portion of the deposit overlying an excavation and left in place as a pillar.
顶柱：位于采场上部的一部分矿体，原地留作矿柱。

5) dip：angle of inclination of a deposit, measured from the horizontal.
倾角：矿床的倾斜角度，即矿体与水平面的夹角。

6) floor：bottom or underlying surface of an underground excavation.
底板：采场底部。

7) footwall：wall rock under the deposit.
下盘：矿体下部岩体。

8) gob：broken, caved, and mined-out portion of the deposit.
采空区：矿床爆破，崩落，出矿后留下的空间。

9) hanging wall：wall rock above a deposit.
上盘：矿床上部岩体。

10) pillar：unmined portion of the deposit, providing support to the roof or hanging wall.
矿柱：未开采的矿体部分，用于支撑采场顶板或上盘岩体。

11) rib：side wall of an excavation; also rib pillar.
间柱：采场侧面的矿柱。

12) roof：back, top, or overlying surface of an excavation.
顶板：地下采场空间的顶部。

13) sill pillar：portion of the deposit underlying an excavation and left in place as a pillar.
底柱：位于采场下部的一部分矿体，原地留作矿柱。

14) strike：horizontal bearing of a tabular deposit at its surface intersection.
走向：扁平矿体在地表水平投影的方向。

15) overhand：advancing in an upward direction.
上向式开采：从下往上进行的开采方式。

16）underhand：advancing in a downward direction.
下向式开采：从上往下进行的开采方式。

［4］Excavation terms 采矿术语：

1）bell：funnel-shaped excavation formed at the top of a raise to move bulk material by gravity from a stope to a drawpoint.
漏斗：把一个天井上部刷大，形成的漏斗状的空间，用于把采场爆破下的矿石依靠自重溜放到出矿口。

2）bleeder：exhaust ventilation lateral.
回风巷道：废气通风的辅助巷道。

3）chute：opening from a drawpoint，utilizing gravity flow to direct bulk material from a bell or orepass to load a conveyance.
溜槽：从出矿口下面的槽，崩落后的矿石利用自重从漏斗或矿石溜井溜放到运输工具之中。

4）crosscut：tertiary horizontal openings，often connecting drifts，entries，or rooms；oriented perpendicularly to the strike of a pitching deposit；also breakthrough.
切割横巷：第三类的地下采矿巷道，切割巷通常与平巷，进路或矿房相连接，它与急倾斜矿床的走向垂直。

5）decline：secondary inclined opening，driven downward to connect levels，sometimes on the top of the dip of a deposit；also declined shaft.
斜井：属于第二类的倾斜的巷道，从上往下开掘的巷道，连接上下水平中段巷道，有时斜井是从矿床顶部开始往下掘进斜井。

6）drawpoint：loading point beneath a stope，utilizing gravity to move bulk material downward and into a conveyance，by a chute or loading machine，also boxhole.
出矿口：采场下部的装载点，矿石利用自重通过溜槽或装载机械溜至运输工具之中。

7）drift：primary or secondary horizontal or near-horizontal opening，oriented parallel to the strike of a pitching deposit.
平巷：相当于第一类或第二类的开拓巷道，水平或近似水平，与矿体走向平行。

8）finger raise：vertical or near-vertical opening used to transfer bulk material from a stope to a drawpoint，often an interconnected set of raises.
指状溜井：垂直或近似垂直的溜井，把崩落矿石从采场溜送至出矿口，通常指状溜井是相互连接的一套溜井。

9）grizzly：coarse screening or scalping device that prevents oversized bulk material from entering a material transfer system；constructed of rails，bars，beams，etc.
格筛：防止大块崩落矿石进入转运系统的网格设备，通常由铁轨，铁棍，横梁制作的网格。

10）haulage way：horizontal opening used primarily for materials handling.
运输巷：主要用于矿岩运输的水平巷道。

11）incline：secondary inclined opening，driven upward to connect levels，also inclined shaft.

斜井：倾斜巷道，从下往上掘进的连接上下水平的倾斜巷道。

12）lateral：secondary or tertiary horizontal opening，often parallel or at an angle to a haulage way，usually to provide ventilation or some auxiliary service.

辅助巷道：属第二类或第三类巷道，辅助巷通常与运输巷平行或成一定角度，同时兼做通风和辅助作业之用。

13）loading pocket：transfer point at a shaft where bulk material is loaded by bin，hopper，and chute into a skip.

装载硐室：在竖井中转运矿石的硐室，在装载硐中矿石通过矿仓，溜槽进入箕斗。

14）longwall：horizontal exploitation opening several hundred feet in length，usually in a tabular deposit.

长臂式采矿法：通常在水平矿体中开采工作面长度达几百英尺的采矿方法。

15）manway：compartment of a raise or a vertical or near－vertical opening intended for personnel travel between two levels.

人行井：在水平中段之间开掘的天井或近似垂直的井，供人员通行之用。

16）orepass：vertical or near-vertical opening through which bulk material flows by gravity.

矿石溜井：垂直或近似垂直的井，用于矿石自溜。

17）raise：secondary or tertiary vertical or near-vertical opening，driven upward from one level to another.

天井：属第二类或第三类巷道的垂直或近似垂直的井，从下往上掘进。

18）ramp：secondary or tertiary inclined opening，driven to connect levels，usually in a downward direction，and used for haulage.

斜坡道：属第二类或第三类的倾斜巷道，用于连接上下水平中段，通常从上往下掘进，用于运输。

19）room：horizontal exploitation opening，usually in a bedded deposit.

矿房：水平回采空间，通常用于层状矿体。

20）shaft：primary vertical of near-vertical opening，connecting the surface with underground workings，also vertical shaft.

竖井：属第一类巷道，垂直或近似垂直的巷道，竖井用于连接地下采场与地表。

21）slope：primary inclined opening，usually a shaft，connecting the surface with underground workings.

斜井：属第一类的倾斜巷道，即斜井，它连接地表与地下采场。

22）slot：narrow vertical or inclined opening opening excavated in a deposit at the end or a stope to provide a bench face.

切割槽：矿体中的宽度不大的垂直或倾斜开挖后的空间，位于矿体端部或采场端部，它能提供台阶工作面。

23）stope：large exploitation opening，usually inclined or vertical，but may also be horizontal.

采场：大的采矿空间，通常是倾斜空间或垂直空间，也可以是水平空间。

24）sublevel：secondary or intermediate level between main levels or horizons，usually close to the exploitation area.

分段：中段中间的水平巷道，中段巷道更加靠近采场。

25）transfer point：location in the materials-handling system，either haulage or hoisting，where bulk material is transferred between conveyances.

装载点：矿岩转移运输的设备的地点，或者是水平运输或者是提升运输。

26）tunnel：main horizontal or near-horizontal opening，with access to the surface at both ends.

隧道：主要水平或近似水平巷道，隧道两端都通达地表。

27）undercut：low horizontal opening excavated under a portion of a deposit，usually a stope，to induce breakage and caving of the deposit；also a narrow cut in the face of a mineral deposit to facilitate breakage.

拉底：在矿体下部开凿的一个水平空间，通常也是一个采场，拉底空间可以诱发矿石破碎或矿体崩落；拉底空间同样也是一个在矿体里开掘的一个窄的切割空间便于破碎矿岩。

Unit 4 Factors Which Require A Shaft

竖井开拓所具备的条件

1. Introduction

Factors which require a shaft are as follows: High output; Increased depth to ore bodies; increased ventilation requirement, such as gassy or hot mines; Bad rock mechanics conditions. The first three points have been dealt with in general when discussing declines.

2. Strong water-bearing rocks

The problem here is to seal off the water whilst sinking or driving. Ground support is not a problem so the access could still be either vertical or inclined.

The water can be dealt with by chemical injection, grouting with cement, or freezing. These methods have all been applied at inclinations from vertical to horizontal. The cost is obviously directly proportional to the lineal meter treated so one tries to pass through the affected area as quickly as possible.

There are two subdivisions in this treatment. Chemical injection and grouting are quick processes carried out from within the shaft or decline, working perhaps 15 to 25 meters ahead, therefore lead time is not large and planning for unforeseen difficulties is easier. For freezing a period of several months may be necessary to establish an ice wall, and a special lining will be necessary to exclude the water when the ice wall thaws. The cost and time involved make this a major project and the returns must be high. This usually means a shaft for a mine and outputs of millions of tonnes over a life of twenty or more years. In civil engineering it would have to be a major road or rail tunnel to justify the cost.

3. Weak water-bearing rocks

One hesitates to define the transition from strong to weak rock in terms of physical strength parameters. A better definition of weak rock would be one that deteriorates rapidly in the presence of water. In some cases the grouting or chemical injection to stop large water flows strengthens the rock to a stage where the ground only requires minimal support.

In practice "weak" ground is where the nature of the rock, particularly its broken condition and where deterioration of clayey materials occurs, causes frequent falls of roof and sides. The intensive support needed, and lost production time, make drivage of inclines particularly expensive. The large amount of roof subject to gravity loading causes continual problems. In these circumstances a shaft is easier to drive and more stable. Depth increments about 2m can be easily meshed and lagged or rock-bolted for temporary protection. The gravity load is not high. In a mine shaft with a multi-stage scaffold the permanent concrete lining can be installed while sinking is in progress and the time to penetrate the affected areas safely is usually very much less than in inclined drivages.

4. Loose Surficial Deposits

These may comprise running sands, loose gravel, soft clay, or peat bogs. In these deposits a vertical shaft which gives a balanced hydrostatic loading on the lining is safest and strongest. The methods used to penetrate these deposits are all easiest with vertical entry because they rely on forming a temporary shield.

(1) Piling: Sheet piles are driven to interlock with each other and form a caisson. The ground is excavated inside. If a series of two to three stages is used at decreasing diameter about 15 to 20 meters per stage can be taken, depending on ease of driving piles. If the piles cannot be driven then the ground may not run. However the major problem concerns substances such as glacial till, where erratic boulders will stop a pile.

(2) Drop-shafts: In shallow wet deposits a steel lining, or a composite steel-concrete lining with a cutting shoe, can be built as an annulus on the surface and loaded to penetrate the soft formation, concurrent if necessary with water-jet cutting and excavation of the material inside the shaft. Muds are often kept suspended as a slurry within the drop-shaft to prevent an inrush into the shaft area. Boulders which deflect the cutting shoe can often be a problem. When the shaft hits bed-rock then various procedures may be undertaken to seal the junction. If possible the shaft should be undercut and dropped into the bed rock.

Some form of grouting may be necessary to make a temporary seal at the drop shaft junction while manual excavation is resumed. Compressed-air work may also be a possibility.

(3) Compressed-air sinking: This can be used alone or at the end of a drop shaft access. An air lock must be built over the shaft area, bolted to at least a short shaft collar which is dead-loaded with scantling to avoid it being lifted out of the ground under pressure. A compromise air pressure is introduced into the sinking shaft. The compromise is between reduced working time at increased pressures and the need to hold back incoming water and mud while the shaft is deepened. Excess air will rapidly pipe away outside the shaft casing and has even caused blow-outs and instability. Too little air will result in an inundation. Specialist experience even more than theoretical knowledge is necessary.

In New South Wales (as a useful reference) detail on compressed air work can be found in the NSW Government Regulations under the Scaffolding and lifts Act 1912-1960 which control most civil engineering type excavation work. The table of safe working periods gives (Table 4.1):

Table 4.1 safe working periods table

Equivalent/kPa (lb/in^2)	Working period/h
0~103 (0~0.103)	8
103~172 (0.103~0.172)	6
172~276 (0.172~0.276)	4
276~345 (0.276~0.345)	3
345~379 (0.345~0.379)	2
379~414 (0.379~0.414)	1.5

Node: decompression period is about 2 hours plus 1 hour for observation from 55~60lb/in^2, 1lb/in^2=0.006897MPa。

(4) Loose deposits underground. Water-bearing sands and weak shales or marls can be encountered underground and will flow into shaft cavities unless controlled. Limited distances may be forepoled dependently on pressure, soil fluidity and water content. Chemical grouting is a possibility and cement injection may also be successful. Thick beds, which must have water in them, and which are known in advance (for example Canadian potash deposits) may be frozen to allow for sinking. Drop shafting could also be made to work underground and jacking lining would be easier.

If pressure is not high, then there is no real problem. The major problem occurs where pressure will force the material into the bottom of the shaft, making access difficult. Techniques are complicated and will require the services of a specialized shaft-sinking company.

Vocabulary

hopper [ˈhɒpə(r)] *n.* 跳跃者；漏斗
portal [pɔːtl] *n.* 大门，入口，正门
inclination [ˌinkliˈneiʃən] *n.* 爱好；倾向；倾斜度
hydrostatic [ˌhaidrəuˈstætik] *adj.* 流体静力（学）的，静水压的
lining [ˈlainiŋ] *n.* 衬里；衬套；隔板
drivage [ˈdraividʒ] *n.* 掘进
lagged [lægd] *v.* 走得极慢，落后；给（管道等）加防冻保暖层
scaffold [ˈskæfəuld] *n.* 脚手架；建筑架
vt. 用支架支撑
surficial [sɜːˈfɪʃəl] *adj.* 表面的；地面的
piling [ˈpaɪlɪŋ] *n.* 桩；桩结构；打桩，打桩工程
caisson [ˈkeisn] *n.* 弹药箱；沉箱

annulus [ˈænjələs]　　n. 环形物
collar [ˈkɒlə(r)]　　n. 环管
　　vt. 控制；捉住
marl [mɑːl]　　n. 石灰泥
　　vt. 施泥灰于
forepole [ˈfɔːpəul]　　n. 超前梁；超前支架
grout [graʊt]　　n. 灰浆
　　vt. 用薄泥浆填塞
scantling [ˈskæntlɪŋ]　　n. 少量；放置木桶的台架
inundation [ˌɪnʌnˈdeɪʃən]　　n. 淹水，洪水，涌到
inrush [ˈɪnrʌʃ]　　n. 侵入；涌入
seal off　　封闭
ease of driving piles　　易于打桩
peat bogs　　泥炭沼泽
sheet pile　　板桩
glacial till　　冰碛
cutting shoe　　切割器
drop-shaft　　沉井
air lock　　气闸
seal the junction　　封住路口

NOTES

[1] Factors which require a shaft are as follows: High output; Increased depth to ore bodies; increased ventilation requirement, such as gassy or hot mines; Bad rock mechanics conditions. The first three points have been dealt with in general when discussing declines.

选择竖井开拓矿床的因素如下：矿山生产能力大；矿床埋藏深；通风需求大，例如瓦斯矿或高温矿井；岩石的力学性能差。前三个因素在讨论斜井时已经谈到过。

[2] The problem here is to seal off the water whilst sinking or driving. Ground support is not a problem so the access could still be either vertical or inclined.

在掘进井筒时要确保封闭水流，不让裂岩层中的水流入井筒。掘井时的地压支护不是大问题，而主要考虑的问题是选择采用竖井开拓还是斜井开拓。

[3] The water can be dealt with by chemical injection, grouting with cement, or freezing. These methods have all been applied at inclinations from vertical to horizontal. The cost is obviously directly proportional to the lineal meter treated so one tries to pass through the affected area as quickly as possible.

通过化学注浆，水泥灌浆或冷冻法可以解决水的问题。根据实际情况，这些堵水的方

法在竖井掘进和斜井掘进中都有应用。很显然，堵水成本与堵水井筒的长度成正比，因此，掘进人员都希望尽量迅速通过这些含水岩层。

[4] There are two subdivisions in this treatment. Chemical injection and grouting are quick processes carried out from within the shaft or decline, working perhaps 15 to 25 meters ahead, therefore lead time is not large and planning for unforeseen difficulties is easier. For freezing a period of several months may be necessary to establish an ice wall, and a special lining will be necessary to exclude the water when the ice wall thaws. The cost and time involved make this a major project and the returns must be high. This usual means a shaft for a mine and outputs of millions of tonnes over a life of twenty or more years. In civil engineering it would have to be a major road or rail tunnel to justify the cost.

堵水有两种细分方法。化学注浆堵水和水泥灌浆堵水是掘进竖井及斜井时最常用的堵水方法，往往堵水要超前掘进 15~25m，因此，超前时间并不算长，处理不可预知的问题就比较容易。采用冷冻法掘进井筒，需要几个月的超前期来形成冰墙，且当冰墙融化时，需要一种特殊衬砌来排除水。冷冻法堵水的时间长，成本高，适用于回报高的大型工程。通常一个矿山的年产量达到几百万吨，矿山开采期达几十年以上的矿山可以采用这种方法掘进井筒。在土木建设工程中的大型公路，铁路隧道中使用冷冻法掘进隧道才能收回成本。

[5] One hesitates to define the transition from strong to weak rock in terms of physical strength parameters. A better definition of weak rock would be one that deteriorates rapidly in the presence of water. In some cases the grouting or chemical injection to stop large water flows strengthens the rock to a stage where the ground only requires minimal support.

人们不容易根据岩石的物理力学性能来区分岩石高或低。当岩石遇水时会迅速破坏，则这种岩石可以定义为岩石强度低。在某些情况下，采用水泥灌浆或化学注浆堵住大部分的水，从而提高岩石强度，因此井巷工程所需的支护强度就最低。

[6] In practice "weak" ground is where the nature of the rock, particularly its broken condition and where deterioration of clayey materials occurs, causes frequent falls of roof and sides. The intensive support needed, and lost production time, make drivage of inclines particularly expensive. The large amount of roof subject to gravity loading causes continual problems. In these circumstances a shaft is easier to drive and more stable. Depth increments about 2m can be easily meshed and lagged or rock-bolted for temporary protection. The gravity load is not high. In a mine shaft with a multi-stage scaffold the permanent concrete lining can be installed while sinking is in progress and the time to penetrate the affected areas safely is usually very much less than in inclined drivages.

在实际工程中，低强度岩石是指特别容易破碎，且含有黏土物质的容易破碎的岩石。这种岩石通常引起采场顶板和侧墙的冒落。这种岩石需要大量支护，而且影响生产能力，在这种岩石中掘进斜井的成本特别高。在这样的岩石中，采场顶板承受上覆岩层重力非常大，会引起一系列问题。在这种情况下，采用竖井开拓就可以更加方便掘进竖井，井筒也更加稳定。每掘进 2m 深就可以很容易地进行锚网临时支护。由于竖井中的重力应力不高，因此，在矿山竖井施工过程中，采用多级支架支护的同时，进行混凝土衬砌永久支护，同时还可以进行掘井作业。在竖井掘进中，通过软弱岩层所需要的时间比斜井要小得多。

Unit 5 Location of Access Relative to Orebodies

主井位置选择

1. Introduction

It may appear rather obvious to state that all mine accesses have two ends, a top and a bottom. However, the point is made to remind readers that both ends have to considered. For an inclined access in a metalliferous mine there is really no problem because an offset distance between top and bottom is easily arranged. For a coal mine drift the problem is not so simple unless the drift starts from the outcrop. The greatest difficulty is with shafts.

2. Surface location

Briefly the requirements for surface location are:

(1) Sufficient flat land, or easy cut-and-fill contours, to establish the mine surface complex. The area has to be calculated after design but the size range without housing could well be 10 to 100 hectares. The latter figure is where large rail-out facilities and stockpile are needed. One can adjust layouts within reason from square to long and thin;

(2) As close as possible to railway, roads and power lines;

(3) Adequate safe disposal areas close by for mullock and tailing;

(4) Adequate water supply, or a site for a dam;

(5) Preferably easy natural pondage on site for process water and for effluent treatment;

(6) Labour supply, preferable with housing. On a coal mine lease of about 8km×8km there is considerable scope for locating the surface plant close to an existing town;

(7) Possible future provision of secondary treatment plants (smelters or coke ovens) which require transport access.

3. Positive choices

(1) Plant needing protective pillars should be sited above the worst geological area (faults, thin coal, dykes, etc);

(2) Bearing in mind any transport contradiction underground, put the pillar over the shallowest part of the seam or orebody because then the protective pillar is smallest and the shaft or decline is also shorter;

(3) Put the surface plant outside the outcrop area or over the footwall of the orebody outside the angle of draw;

(4) Avoid placing any surface plant close to the mine. Costs may be such that if material is belt-hauled or truck-hauled off the lease to a central process plant then the total running cost of a complex could be lower. A conveyor belt is not affected by subsidence;

(5) Shafts can have jacking point under the headgear and buildings can be flexible. Between 40% and 100% of coal or mineral can then be extracted from under the surface structure that must be close by , for example the change house.

4. Avoid

(1) Bad geological areas where the sinking would be difficult;

(2) Low-lying areas susceptible to infrequent but periodical flooding.

5. Underground location

General requirements: Remember that the bottom of an access doesn't have to be immediately under the top. Although modern technology has largely obviated the need for multi-stage hoisting, and inclined and compound shafts (bent ones) are obsolescent, there can be good technical reasons why their economic short-comings should be accepted.

Drifts and inclined spiral or zig-zags give good scope for adjustment to the best of both worlds-the surface and the underworld. A 1 in 3 drift gives about 1000m on 300m with transition curves and turn-outs or bunkers. A pair of declines at 1 in 9 could give mining access across a whole leases with through ventilation.

The concept of centre of gravity. It has been frequently suggested that a shaft should be sited at the centre of gravity of an orebody. This concept should be examined carefully before it is accepted too readily. The basic premise is that the cost of haulage of development rock and of mineral should be minimized. Mathematically, where H equals haulage costs and X is a set of factors that includes haulage distance, method of transport, gradients, environmental effects on transport, number of transfer points on haulages, and so on.

It is a gross over-simplification to use some of the older formula of the type that attempt to balance quantities on each crosscut or level from left and right of a shaft by interactive processes. The form was: where q and d are quantities and distances on the left and right respectively. To balance accurately the levels must be of the same inclination sense (that is both flat or both at the same angle up or down) and have the same number of crosscuts, ventilation doors, etc., as well as the same transport system. These theories were developed for symmetrically shaped coal seam deposits in the days of hand-working and rail transport.

Vocabulary

offset [ˈɒfset] *n.* 分支；补偿

	vt. 抵消
dykes [daɪks]	*n*. 堤；坝；沟
lease [liːs]	*n*. 租借
	vt. 出租
subsidence [səbˈsaɪdns]	*n*. 沉淀；沉下
obviate [ˈɔbvieit]	*vt*. 排除；避免
jacking [ˈdʒækɪŋ]	*n*. 打千斤，用千斤顶打桩
headgear [ˈhedgɪə(r)]	*n*. 帽子；头上戴的东西
offset distance	偏移距离
coke oven	焦炉
protective pillar	保安矿柱
turn-outs	转弯
zig-zag	之字形

NOTES

[1] It may appear rather obvious to state that all mine accesses have two ends, a top and a bottom. However, the point is made to remind readers that both ends have to considered. For an inclined access in a metalliferous mine there is really no problem because an offset distance between top and bottom is easily arranged. For a coal mine drift the problem is not so simple unless the drift starts from the outcrop. The greatest difficulty is with shafts.

很明显所有矿山井筒都有两个端口：一个顶部端口，一个底部端口。两个端口都得考虑周全。对于金属矿山的斜井而言，顶底部之间的偏移距离可以很容易进行调整，因此斜井不存在实际问题。对于煤矿的平硐而言，当平硐不是从露头开始掘进，平硐掘进的问题就不是那么简单了。竖井开拓碰到的困难最多。

[2] Briefly the requirements for surface location are:

(1) Sufficient flat land, or easy cut-and-fill contours, to establish the mine surface complex. The area has to be calculated after design but the size range without housing could well be 10 to 100 hectares. The latter figure is where large rail-out facilities and stockpile are needed. One can adjust layouts within reason from square to long and thin;

(2) As close as possible to railway, roads and power lines;

(3) Adequate safe disposal areas close by for mullock and tailing;

(4) Adequate water supply, or a site for a dam;

(5) Preferably easy natural pondage on site for process water and for effluent treatment;

(6) Labour supply, preferable with housing. On a coal mine lease of about 8km×8km there is considerable scope for locating the surface plant close to an existing town;

(7) Possible future provision of secondary treatmentplants (smelters or coke ovens) which require transport access.

简言之，工业广场在地表位置的选择应满足如下要求：

（1）地表具有足够的平整土地，或采用简单的挖方或填方就可以平整土地，可以布置矿山工业设施。地表区域大小要满足设计要求。如果地表不设住房，则地表面积为 10~100hm² 就可以满足要求。当地表面积达 100hm² 时，地表可以布置铁路运输线路和矿仓。地表区域可以根据具体情况采用正方形或长方形；

（2）工业广场尽量布置在靠近铁路、公路或动力线的地方；

（3）工业广场附近有足够的容纳废石和尾砂的安全场地；

（4）工业广场附近具有足够的水源或能建设堤坝储存水；

（5）工业广场附近最好有天然水池为选矿服务和污水处理；

（6）工业广场附近有足够的劳动力，有住房。例如，某个煤矿在靠近一个现有城镇的地方租用了 64km² 的地表面积用于布置矿山相应的设施；

（7）还要考虑将来矿山扩建所需要的土地面积（如增加冶炼或炼焦）及所需要的运输线路的面积。

Unit 6　Mine development of Large Metalliferous Orebodies

大型金属矿床的地下开拓

1. Introduction

These orebodies will certainly require shafts and levels and probably declines for access for machinery between levels. If one has to opt for a modern development as typical, then Cobar Mines is probably very representative. Its benefits have been such to the company that before the metal price crisis of the mid-1970s Cobar had deepened both shafts and added a third major haulage level.

2. Mine Developments

Shafts: The location of these has been dealt with earlier.

Levels: The interval between levels depends on: (1) Geology and size of orebody; (2) Strength of backs and walls; (3) Mining methods; (4) Life of ore passes; (5) Capital available for equipment on levels, and for driving between levels; (6) Access for men; (7) Current technology; (8) Surveying convenience.

(1) Geology and size of orebody. A badly faulted and disturbed orebody will need many exploration levels and diamond drill holes. Variable grade bodies will also need more exploration, and probably will need selective mining. Consequently a large orebody which really consists of many scattered lenses of variable grade will have to be developed with a close level interval (say 20 to 30m) and mined selectively.

Size of orebody will control levels on a simple multiple basis. If the orebody is 400m deep, even though a 180m level interval is feasible, the orebody is too deep for two levels, so it may be conveniently split into three, for instance two at 130m and the bottom at 140m.

(2) Strength of backs and walls and mining methods. Because these two are so closely related, indeed 2 controls 3, they should be considered together. A strong ore and strong country rock enable large open stopes to be worked, or large hanging wall area to be left open. Other things being equal, stopes should be worked as high as possible because that reduce the cost of drawpoints and chutes to a minimum for the mine. Cut and fill methods enable large intervals to be obtained between levels, because ring drilling is unnecessary withing stopes.

If current development in down-the hole drilling of 100mm diameter and larger holes is discounted, then drilling technology at present limits straight holes to about 20m, because ring-drilled holes must over-lap about 1.5 to 2m to obtain good fragmentation, then there must be at least a drill sub-level every 36m, say 35m to round it off. Some mines have experimented with longhole open stope about 40 to 60m high, drilled up and down from levels about 80 to 120 m apart. However these mines have found that there is a tendency for a bridge of ore to develop at the mid-line of the stope which is then difficult to break out.

As ores and rocks becomes weaker, closer support methods are adopted until eventually one considers caving. In sub-level caving the sub-levels may be only 15 to 18m apart, so that there is a tendency towards closer spacing even of main levels, although this is offset by ore pass development.

Block caving tends to increase the spacing between levels because access is not normally needed into the stopes and a level interval of 150 to 180m is easily attainable, indeed it is probably beneficial to assist development of a good cave.

(3) Life of ore passes. In most stoping methods ore is drawn off through an ore pass at some stage. The pass may be through ore, through rock, or may be a lined pass through fill. All of these have a finite life after which they are so enlarged as to be unsafe, with subsequent cave-ins, or slabbing of the walls may cause heavy dilution, followed by a cave-in.

Once the life of an ore pass has been determined in tonnes, the cost of spacing x length can be determined. Obviously the fewer the ore passes in relation to tramming runs, the lower the cost. However if there are too few ore passes they will wear out and limit the stope height and hence decrease the level interval.

Dependent on the size of the mine a manual or computer program must be carried out to minimize the values (ore pass interval x ore pass height x ore pass lining costs x cost per level driven x tramming costs to ore pass x trucking costs per loading chute) for the whole mine. To qualify all these factors is so complex that there is a tendency to treat ore pass life as a simple problem of increasing the number of passes, or of changing the mining method.

In practice, if ore passes are too weak in ore they are put into the rock. If the rock is too soft for a good ore pass, the whole mining operation will be difficult and expensive, and level intervals will be close.

(4) Capital available. If capital is difficult toraise, small equipment will be used and operating costs tend to be high. Small equipment usually needs more men and easy access up ladders to stope, consequently levels intervals are close. However shaft and skip size will be small and loading pocket construction will not be expensive.

High capital expenditure has to be justified by large outputs. This needs automated winding arrangements and expensive plat (station) installations.

It also needs big reliable locomotives, large mine cars, heavy rails, good track, etc. So that the capital cost of equipment per level is high. One obviously must keep haulage levels to a minimum. However ore pass life and access to stopes can be a problem so that one may compromise with a major haulage level at (say) 150 to 450m intervals, and access levels for men and materials at 75 to 150m intervals between these major levels.

(5) Access for men. In most stoping methods men must have easy access. If they have to climb ladders then about 50m may be a limit. Modern technology may compromise by adding in the cost of service cages, or even lifts, to take men up through an access raise into the stopes. With an increasing use of raise borers to improve level intervals, a raise-bored "lift" shaft becomes an attractive possibility.

(6) Current mines technology. Older mines were limited by hand-raising techniques to short intervals between levels. A study of mine plans will show that. As machines such as the Alimak raise climber, and techniques such as long-hole raise blasting, were tried, the interval between levels gradually increased from 15 to 20m, then 30m and finally to about 50m. Now that raise boring is more common in large mines, and cutter technology and hydraulic thrust equipment has been improved, raise-bored holes of 300m or more between levels are feasible. This means that a development such as Cobar Mines, with major levels at 180m intervals and a decline access ramp for mining, is currently the most economic. The fewer levels there are, the more money can be spent on them and on ore pass development.

The limitation of ring drilling technology has been mentioned above. In sublevel open stoping small blastholes (about 50mm) tend to deviate at depths of about 15m and may be 0.3m out of line at 20m into the hole. With spacing and burdens of the order of 1.2 to 1.5m, a deviation of 0.3m each on adjacent holes may be a 50% increase in spacing. This may cause either poor fragmentation or a complete failure to break a ring, with failure of subsequent rings in the same blast.

A current development to extend the range of long-hole open stoping is to emulate surface practice. Down-the-hole hammer drills with bit diameters of 100 to 150mm are being used. These can drill 20 to 30m without noticeable deviation because the impact is at the bottom of the hole, which reduces rod whip. Hole lengths of 50 to 60m are possible and large decked charges can be used so the possibility of poor fragmentation is reduced. The limitations would appear to be increased blast vibration, possibly more fumes underground, and because burden and spacing are increased, a greater likelihood of oversize boulders needing secondary blasting.

These larger boulders may be in a stope that cannot be entered so that drawpoints will become blocked. If one increases the power factor to ensure good fragmentation then increased powder costs offset the reduced drilling costs, and there will also be more blast vibration. However, if longhole blast stopes can be increased in height there is comparable technology to increase the level interval.

(7) Surveying convenience. After all the previous factors have been developed it would be normal practice to round the level interval out to a convenient figure. For instance it would be unlikely that a theoretical figure of 187.6m would be adopted, and the planning engineer would either drop back to 180m or round up to 200m, to allow easy measurement, and easier construction of mine plans.

(8) Ore pass development. Improvement in raise-driving techniques have enabled ore passed to

be driven over greater intervals. Contemporary wage increase have encouraged mines to keep-non-productive employment, such as haulage work, to a minimum. Near-vertical ore passes in rock can provide a means of concentrating output from several stopes into a central ore bin close to the shaft. The ore or mullock, can be run through a primary crusher and onto a conveyor to fill the skip pockets for automated hoisting procedures.

The principal requirement is for a large orebody with sufficient vertical height to make the development economically worthwhile, and to permit gravity concentration of output at one point. An economic alternative can be develop ore passed over two or three levels with open drawpoints and use LHDs or even front-end-loaders to tram the ore to a second group of ore passes.

Yet a further arrangement adopted at Mt Isa was to make every second level at approximately 110m intervals a haulage level with chutes and locomotive haulage. Ore from intermediate levels was fed through ore passes, but the plan area of the mine is extremely large. The ore pass system was developed for the central portion.

3. Future development

Ore pass development as a logical full scale system would have ore and mullock passes developed over heights of 300m or more in the footwall of the orebody.

This would imply future developments as:

(1) A primary and secondary crusher installed at the bottom of the mine, or approximately 450m intervals, with a conveyor to load (over any distance) into an ore bin/skip pocket system at the shaft;

(2) Major levels at about 300m or 450m intervals;

(3) Cage access to intermediate level for drilling, and stope access through raise bored shafts; or for smaller orebodies levels at about 100m intervals in the main shaft and a service cage system between levels.

Under these circumstances the level interval becomes a surveying exercise in studying ore pass angles and dividing the orebody into neat geometrical patterns. After the strategic development plan has been drawn up, only minor relocation should be necessary to suit local geological and rock mechanics difficulties.

Vocabulary

opt [ɑ:pt] *vi.* 选择；决定

grade [greid] *n.* 品位
vt. 减少坡度；逐渐变化

scattered [ˈskætəd] *adj.* 散乱的；分散的

level [ˈlevl] *n.* 水平（仪）；标准；程度［矿物］主平巷

discount [ˈdɪskaʊnt] *v. n.* 折扣；贴现；预期
hanging wall 顶板
sub-level 分段巷道
mid-line 中线
ore pass 矿石溜井
a lined pass through fill 充填体内顺路溜井（内衬溜井）
tramming runs 电机车运输
longhole blast 深孔爆破
access raise 人行天井
bridge of ore 采场悬拱

NOTES

[1] These orebodies will certainly require shafts and levels and probably declines for access for machinery between levels. If one has to opt for a modern development as typical, then Cobar Mines is probably very representative. Its benefits have been such to the company that before the metal price crisis of the mid-1970s Cobar had deepened both shafts and added a third major haulage level.

大型金属矿床要采用竖井和水平巷道进行开拓，同时在中段之间还可能开掘斜坡道方便机械设备上下转移。卡巴矿就是一个典型的现代矿山竖井开拓的实例。在 20 世纪 70 年代中期金属价格低迷之前，该矿就已经延伸的竖井，并增设了第三个主运输中段，从而使得矿山收益。

[2] Shafts: The location of these has been dealt with earlier.

Levels: The interval between levels depends on: (1) Geology and size of orebody; (2) Strength of backs and walls; (3) Mining methods; (4) Life of ore passes; (5) Capital available for equipment on levels, and for driving between levels; (6) Access for men; (7) Current technology; (8) Surveying convenience.

竖井：竖井位置选择已经进行了叙述。

中段：两中段之间的高度取决于：(1) 矿床地质条件和矿床大小；(2) 矿床顶、底板和侧面岩石强度；(3) 采矿方法；(4) 矿石溜井使用期限；(5) 购买中段运输设备以及采场设备的资金量；(6) 人员通道；(7) 当前的技术水平；(8) 便于勘探。

[3] A badly faulted and disturbed orebody will need many exploration levels and diamond drill holes. Variable grade bodies will also need more exploration, and probably will need selective mining. Consequently a large orebody which really consists of many scattered lenses of variable grade will have to be developed with a close levelinterval (say 20 to 30m) and mined selectively.

矿体被断层带错动的矿体需要在多个水平进行勘探和钻探来探明矿体。当矿体品位分布不均匀时，也需要更多的勘探工作，并采取不同的采矿方法。因此，当矿体品位分布不均匀，且由多个分散的透镜状矿体组成的大型矿体的开拓水平之间的高度要降低（如 20~30m），而且要进行选择性开采。

[4] Size of orebody will control levels on a simple multiple basis. If the orebody is 400m deep, even though a 180m level interval is feasible the orebody is too deep for two levels, so it may be conveniently split into three, for instance two at 130m and the bottom at 140m.

根据开拓水平高度和矿体延伸尺寸就可以确定矿床开拓水平数量。如果矿体垂直延伸 400m，尽管矿床开拓可以采用阶段高度 180m，那么设计两个中段进行开拓就不能满足矿床开拓的需要，因此，需要设计成三个中段进行开拓，即设计两个 130m 高的中段，最下一个中段采用 140m 高。

[5] Because these two are so closely related, indeed 2 controls 3, they should be considered together. A strong ore and strong country rock enable large open stopes to be worked, or large hanging wall area to be left open. Other things being equal, stopes should be worked as high as possible because that reduce the cost of drawpoints and chutes to a minimum for the mine. Cut and fill methods enable large intervals to be obtained between levels, because ring drilling is unnecessary withing stopes.

因为矿床顶底板及侧面岩石强度与采矿方法这两个因素密切相关，其实，矿床顶底板及侧面岩石强度性能是采矿方法的选择重要因素，因此，矿床顶底板及侧面岩石强度与采矿方法必须一起考虑。矿石与围岩强度大，则采场空间就大，采空区的顶板面积就大。同样，采场高度也应该尽量大一些，因为采场的高度高，可以减少掘进出矿口和溜槽数量，从而降低成本。采用充填法采矿，可以增加中段高度，因为充填采矿法的采场内无须钻凿环形炮孔。

[6] If current development in down-the hole drilling of 100mm diameter and larger holes is discounted, then drilling technology at present limits straight holes to about 20m, because ring-drilled holes must over-lap about 1.5 to 2m to obtain good fragmentation, then there must be at least a drill sub-level every 36m, say 35m to round it off. Some mines have experimented with longhole open stope about 40 to 60m high, drilled up and down from levels about 80 to 120m apart. However these mines have found that there is a tendency for a bridge of ore to develop at the mid-line of the stope which is then difficult to break out.

如果 100mm 或直径更大的下向钻机没有进展，那么目前的钻孔技术就只能钻凿孔深 20m 的直孔。因为环形炮孔之间需要相互重叠约 1.5~2m 才能获得良好的破碎效果，因此，需要每隔 36m 设置一个分段凿岩巷，按分段高度取 35m 计算，那么就可以使环形炮孔重叠起来。有些矿山采用过深空凿岩的空场法，其孔深达 40~60m，从中段凿岩巷道中分别向上和向下凿岩，中段高度约 80~120m。可是，这些矿山的实践表明，在采场中部若出现矿石拱顶的趋势，处理起来比较困难。

[7] As ores and rocks becomes weaker, closer support methods are adopted until eventually

one considers caving. In sub-level caving the sub-levels may be only 15 to 18m apart, so that there is a tendency towards closer spacing even of main levels, although this is offset by ore pass development.

当矿石与围岩变得软弱时，采场支架需要更加密集，如果矿岩更加软弱，则应采用崩落法回采。分段崩落法的分段高度可以为 15~18m，因此其发展趋势是中段高度降低，但掘井的溜井数量增加。

[8] Block caving tends to increase the spacing between levels because access is not normally needed into the stopes and a level interval of 150 to 180m is easily attainable, indeed it is probably beneficial to assist development of a good cave.

阶段崩落法的发展趋势是增加中段高度，因为无须掘进巷道进入采场，其中段高度可以达到 150~180m，这个中段高度其实是有益于矿体获得更好的崩落条件。

Unit 7 Underground Metalliferous Mining

金属矿床地下开采

1. Introduction

This chapter will deal with orebodies that have significant vertical dimensions. It is essentially three-dimensional mining in which lateral development roadways are driven at successive depths and connected vertically. Mining then proceeds upwards in overhand stoping, or downwards in underhand stoping. Steeply inclined thin orebodies (veins) or massive bodies are worked on these general methods.

A change in some thin bodies from working along the strike to working down dip is made when the footwall dips at less than approximately 45°. At about this angle the ore will no longer flow under gravity and has to be moved mechanically. As an orebody becomes substantially flat and is less than a few meters thick then it may be worked as a bedded deposit as described.

Breast stoping is a term sometimes used for the room and pillar methods or for slot mining.

Only general methods of mining will be described. Current technical journals publish many interesting variations that are devised to meet local conditions. Techniques have been basically similar for many years but improvement in machinery, and a chronic shortage of mining personnel have made intensive mechanisation a prerequisite of most mining operations. This has changed the emphasis and application of techniques, and has altered many traditionally held viewpoints of the relative advantages and disadvantages of methods.

As the older generation of skilled craftsmen retire, and the richer remnants of ore disappear, it is likely that some methods such as square set stoping and top slicing may have to disappear with them. Small mines with a handful of men and low overheads may still find them profitable.

2. Classification of stoping methods

Ores are won by stoping, which is a general term used to cover any mining method. The basic classification of methods devised by the U. S. Bureau of Mines in 1936 is still valid, the names are the same, but techniques and relative importance have altered, and some additions must be made. This classification is based on a transition from strong rocks and ore to weak rocks and ore. These are the primary factors in a choice of mining method. In general the larger the opening that can be left unsupported, then the greater the degree of mechanization that can be adopted, and consequently the greater the safety and economy of the operation.

Mineral exploitation in which all extraction operations are carried out beneath the earth's surface is termed underground mining. Underground methods are employed when the depth of the deposit, the stripping ratio of overburden to ore (or coal or stone) or both become excessive for surface exploitation. Once the economics has been established, then the selection of a proper mining method hinges mainly on: (1) determining the appropriate form of ground support, if necessary, or its absence; (2) designing an appropriate configuration and extraction sequence to conform to the spatial characteristics of the mineral deposit.

Reflecting the importance of ground support, underground mining methods are categorized in three classes on the basis of the extent of support utilized. They are unsupported, supported and caving. The unsupported class, the subject of this chapter, consists of those underground methods which are essentially self-supporting and require no major artificial system of support to carry the superincumbent load, relying instead on the walls of the openings and natural pillars. The superincumbent load is comprised of the weight of the overburden and any tectonic forces acting at depth. This definition of unsupported methods does not preclude the use of rock or roof bolts or light structural sets of timber or steel, provided that such artificial support does not significantly alter the load-carrying ability of the natural structure.

Theoretically, the unsupported class of methods can be used in any type of mineral deposit (except placer) by varying the ratio of span of opening to width of pillar to achieve the desired mine life expectancy. Since the stable size of opening is determined by the depth and the strength properties of the ore and overlying rock, the safe span conceivable could range from a few feet (meters) to over 100ft (30m). Practically, the unsupported methods are not universally applicable and are limited to deposits with favorable characteristics. The unsupported class, however, is still the most widely used underground, accounting for over 80% of the U. S. mineral production from subsurface mines.

When exploitation workings cannot be held open for the required life expectancy without the substantial use of artificial support systems, supported methods are used. When the deposit and overlying rock are sufficiently weak and subsidence is tolerable, then support is withheld. Undercutting is carried out as necessary and a caving method is employed. Supported methods are little used today, while caving is growing in popularity.

Others factors, of course, enter into the selection of a method. Their influence is considered in the discussion of each method and also in the summary for underground mining methods, as was done with surface methods. Although method selection remains largely an empirical art, the evaluation of all determining factors must be done as objectively as possible.

While there is a lack of agreement among the various schemes to classify underground methods, we shall employ the generic one presented below. In that classification, the following are considered unsupported methods:

(1) Room and pillar mining.

(2) Stope and pillar mining.

(3) Shrinkage stoping.

(4) Sublevel stoping.

Unlike surface mining, there is little distinction in the cycle of operations for the various underground methods (except in coal mining), the differences occurring in the direction of mining (vertical or horizontal), the ratio of opening-to-pillar dimensions, and the nature of artificial support used, if any. Of the unsupported methods, room and pillar mining and stope and pillar mining employ horizontal openings, low opening-to-pillar ratios, and light-to-moderate support in all openings. Shrinkage and sublevel stoping utilize vertical or steeply inclined openings (and gravity for the flow of bulk material), high opening-to-pillar ratios, and light support mainly in the development openings.

Vocabulary

metalliferous [ˌmetəˈlifərəs]	*adj*. 产金属的；含金属的
stoping [ˈstɒpɪŋ]	*n*. 采场，回采
overhand [ˈəuvəhænd]	*adj*. *adv*. 举手过肩的（地）
underhand [ˈʌndəhænd]	*adj*. 秘密的；卑鄙的
	ad. 秘密地；阴险地，狡诈地
strike [straik]	*vt*. 打；采取；停止
	vi. 打；罢工
	n. 罢工；优良品质
slot [slɔt]	*n*. 狭槽
	vt. 在……开一狭缝
breast [brest]	*n*. 胸膛；[矿物] 工作面
chronic [ˈkrɔnik]	*adj*. 慢性的；长期的
prerequisite [ˈpriːˈrekwizit]	*n*. 先决条件；前提
remnant [ˈremnənt]	*n*. 残余；遗物；剩余物；残留
transition [trænˈziʒən]	*n*. 转变；过渡；迁移；[地理] 转移层
hinge [hindʒ]	*n*. 铰链；重点
	vt. 装以铰链
	vi. 依铰链而转动；依…而定
caving [ˈkeɪvɪŋ]	*n*. 崩落法
superincumbent [sjuːpəinˈkʌmbənt]	*adj*. 在上面的；自上而下的；盖在上面的
overburden [ˌəuvəˈbəːdn]	*vt*. 使负担过重
	n. 过重的负担；[矿物] 覆盖层
tectonic [tekˈtɔnik]	*adj*. 构造的；筑造的；建筑的
generic [dʒiˈnerik]	*adj*. 一般的；普通的
	n. 没有商标的商品
scheme [skiːm]	*n*. 计划；图表；图式
	vi. 拟订计划
	vt. 计划

distinction [dis'tiŋkʃən]	*n.* 区分；特征；差别
lateral development roadway	横向开拓巷道，水平巷道
overhand stoping	上向式采场
underhand stoping	下向式采场
bedded deposit	层状矿床
square set stoping	方框支架采矿法
stripping ratio of overburden to ore	剥采比
spatial characteristics of the mineral deposit	矿体的几何形态
unsupported mining method	无支护采矿法，空场法
supported mining method	有支护采矿法，充填法
superincumbent load	上覆岩层载荷
roof bolt	岩石锚杆
life expectancy	正常开采期间
room and pillar mining	房柱法
stope and pillar mining	全面法
shrinkage stoping	留矿法
sublevel stoping	分段空场法

NOTES

[1] This chapter will deal with orebodies that have significant vertical dimensions. It is essentially three - dimensional mining in which lateral development roadways are driven at successive depths and connected vertically. Mining then proceeds upwards in overhand stoping, or downwards in underhand stoping. Steeply inclined thin orebodies (veins) or massive bodies are worked on these general methods.

本章讲述垂直延伸矿体的开采。在矿床的三维开采空间中，需要在各个中段掘进水平开拓巷道与竖井相通。采场中的采矿作业可以采用上向式开采，也可以采用下向式开采。急倾斜矿体（矿脉）或大型矿体可以采用这样的开采方法。

[2] A change in some thin bodies from working along the strike to working down dip is made when the footwall dips at less than approximately 45°. At about this angle the ore will no longer flow under gravity and has to be moved mechanically. As an orebody becomes substantially flat and is less than a few meters thick then it may be worked as a bedded deposit as described.

当某些薄矿体，其矿体底板倾斜角度小于45°，则作业方式需要调整，即沿矿体走向开采变为沿矿体倾斜方向向下开采。在这种角度条件下，矿石不再靠自重在采场流动，而需要采用机械方式进行搬运。当矿体近似水平时，且矿体只有几米厚，那么，这样的矿床开采方式可以采用水平层状矿体的开采方式进行开采。

[3] Ores are won by stoping, which is a general term used to cover any mining method. This

classification is based on a transition from strong rocks and ore to weak rocks and ore. These are the primary factors in a choice of mining method. In general the larger the opening that can be left unsupported, then the greater the degree of mechanization that can be adopted, and consequently the greater the safety and economy of the operation.

矿石是在采场内进行开采的，所有采矿方法都有采场这一名词。矿岩稳固与矿岩破碎都采用这种分类方法。矿岩稳固性是选择采矿方法的主要依据。通常，无须支护的采场空间越大，则机械化程度越高，相应的生产安全性越好，生产成本越低。

[4] Mineral exploitation in which all extraction operations are carried out beneath the earth's surface is termed underground mining. Underground methods are employed when the depth of the deposit, the stripping ratio of overburden to ore (or coal or stone) or both become excessive for surface exploitation. Once the economics has been established, then the selection of a proper mining method hinges mainly on: (1) determining the appropriate form of ground support, if necessary, or its absence; (2) designing an appropriate configuration and extraction sequence to conform to the spatial characteristics of the mineral deposit.

在地下进行的采矿作业称为地下采矿。当矿床埋藏太深，露天开采时的剥采比太大，这样的矿床就只能采用地下开采。在确定矿床开采经济性后，选择合理的采矿方法主要取决于如下两点：(1) 确定合理的支护方式（如果需要支护）或确定不支护时对应的采矿方法；(2) 根据矿床的赋存条件，设计合理的采场几何尺寸。

[5] Reflecting the importance of ground support, underground mining methods are categorized in three classes on the basis of the extent of support utilized. They are unsupported, supported and caving. The unsupported class, the subject of this chapter, consists of those underground methods which are essentially self-supporting and require no major artificial system of support to carry the superincumbent load, relying instead on the walls of the openings and natural pillars. The superincumbent load is comprised of the weight of the overburden and any tectonic forces acting at depth. This definition of unsupported methods does not preclude the use of rock or roof bolts or light structural sets of timber or steel, provided that such artificial support does not significantly alter the load-carrying ability of the natural structure.

地下开采过程中，地压管理方法非常重要，因此根据所使用的地压管理方法，将地下开采方法分为三大类：即空场法、充填法和崩落法。空场法是本章主要叙述对象，空场法主要包括几种采矿法，这几种空场采矿法基本靠自身留下的矿柱和顶板支撑地压，而不需要大量人工支护措施。采场上部荷载是由上覆岩层自重和构造应力组成。空场法的定义中并没有排除使用岩石或顶板锚杆，轻型木支架或钢支架来支撑地压。只要这些人工支护没有对采场矿柱的承载能力有实质性的改变，则采用矿柱支撑地压的采矿方法都称为空场法。

Unit 8 Room and Pillar Mining

房柱法

1. Introduction

In room and pillar mining, openings are driven orthogonally and at regular intervals in a mineral deposit—usually flat-lying (or nearly so), tabular, and relatively thin—forming rectangular or square pillars for natural support. If the deposit and method are very uniform, the appearance of the mine in plan view is not unlike a checkerboard or the intersecting streets and rectangular blocks of a city. Both development openings (entries) and exploitation openings (rooms) closely resemble one another; both are driven parallel and in multiple, and when connected by crosscuts, pillars are formed. Driving several openings at one time increases production and efficiency by providing numerous working places and improves ventilation and transportation as well.

By its nature, room and pillar mining is ideally suited to the underground production of coal and of several nonmetallic and a few metallic minerals. Largely an American method, it matured with the coal industry, and today still accounts for nearly 85% of U. S. underground coal production and about 62% of U. S. underground mineral production. Properly considered a large-scale method, it faces increasingly stiff competition from long-wall mining, a caving method which is highly productive and responsible for the balance of coal produced below ground. Our discussion here is mainly of the application of the room and pillar method to the mining of coal.

A generalized representation of room and pillar mining appears in Fig. 8.1, comparing conventional and continuous equipment. So-called conventional mining is cyclical, employing mobile, mechanized equipment to carry out the production cycle of operations. As shown, it requires at least 5 working places for a smooth cycle and as many as 8~12 (to allow for delays) for high efficiency. With continuous mining, separate unit operations are eliminated or performed by a single high-performance continuous mining machine. One working place theoretically suffices for the miner with a second for the bolter, but again, for high efficiency, more places (4~6) are provided. Of today's coal production by the room and pillar method, approximately 25% is mined by conventional equipment and 75% by continuous, with the trend toward continuous mining. In general, conventional equipment is advantageous in hard seams, seams with hard rock partings, gassy conditions, and variable seam height; continuous is superior in design as a non-cyclic system and advantageous in thin seams, where there is a bad roof or a limited number of working places, and produce a fine product.

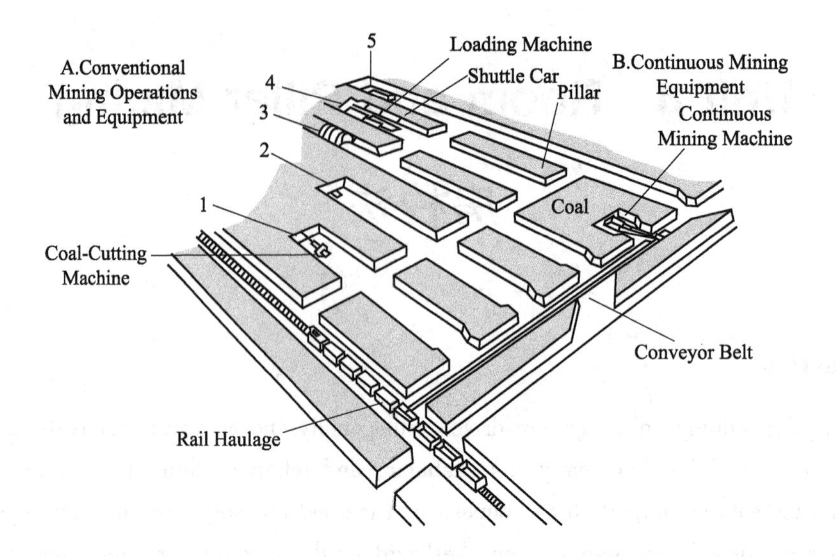

Fig. 8. 1 Generalized representation of room and pillar mining.
Both conventional and continuous equipment are shown
1—Undercut of top-cut coal face, drill holes for explosives; 2—Charge holes; 3—Blast; 4—Load coal; 5—Roof bolt

In addition to number of working places, there are several other design parameters to be selected in room and pillar mining, the most important of which are the dimensions of the openings. The height of opening, of course, is equal to the thickness of the bed or deposit, unless the bed is too thin to permit mechanized operations, in which case some rock in the roof or floor is mined along with the coal (typically, U. S. coal seams that are minable underground are 2. 5 ~ 15ft, or 0. 8 ~ 4. 5m thick). The width ofopening (span) for production or recovery reasons should be a maximum, but for roof control and safety reasons, it is limited by MSHA (to 20ft, or 6m, if roof bolts are used, or to 9m, if other support is used as well). The spacing (centers) of openings, entries or rooms, is sufficiently close that the stress distribution around one opening may affect that around an adjacent openings; it varies from 12 ~ 30m in coal mines. Finally, the maximum spacing between crosscuts is limited by ventilation concerns and is usually specified in state laws.

In design practice, the selection of pillar size (or ration of pillar-to-opening widths) and the nature and amount of support are carefully coordinated. When an opening is created, the unit weight of overburden above it is transferred by an arching action to the sides of the opening or adjacent pillars. Because high stresses result, it may be necessary to design the width of the pillar to avoid stress super-positioning, especially in the case of long-life entries. To do so requires a pillar width equal to at least three times the opening width. For example, if entries are driven 6m in width, their center spacing must be 24m to provide a pillar width of 18m, a ratio of 3 : 1. With such a design, a square roof-bolt pattern on 1. 2 ~ 1. 5m centers would probably be likely. In exploitation, wider rooms (9m) would probably be driven on closer centers (say, 12 ~ 18m) because of their more temporary nature.

The layout of most room and pillar mines is remarkably similar, even under different conditions and with different equipment. Once the main and subsequent development has been completed, production panels of rooms are driven on one or both sides of the room entries. Panels range from 600~1200m in length, and rooms vary from 90~120m in length. Exploitation proceeds either on the advance or the retreat; Fig. 8. 2 illustrates a drift mine, with main and panel entries only, and rooms driven on the retreat. Since pillars are not recovered, rooms are driven a maximum width 9m and pillars a minimum (3m).

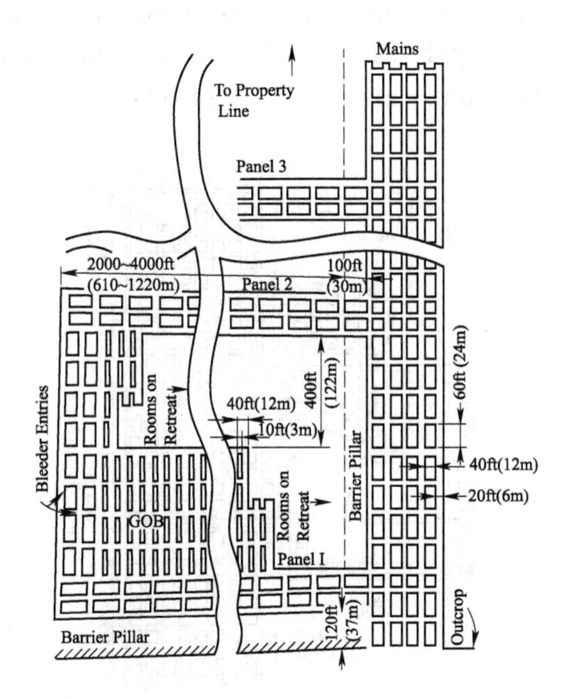

Fig. 8. 2 Room-and-pillar mining, driving rooms on the retreat without pillar recovery

Recovery in room and pillar mining is enhanced by extractingpillars (normally only chain pillars, those formed between adjacent openings are taken; barrier pillars are allowed to remain). This practice is referred to as second mining or pillar recovery (room driving is termed first mining). Partial extraction recovers some pillars or portions of most pillars; wholesale roof collapse and subsidence usually do not occur. If more complete or total extraction is practiced, then caving is induced and surface subsidence eventually occurs. In that event, room and pillar mining ceases to be an unsupported method and becomes a caving method. Among the conditions which should prevail collectively for essentially complete pillar recovery: (1) an immediate roof that requires little support; (2) an overlying series of beds several times the thickness of the deposit which will cave readily; (3) a thick seam of high-quality coal; (4) moderate to deep cover; (5) skilled personnel and supervision.

Fig. 8. 3 depicts a room and pillar coal mine in which complete extraction of pillars (except bleeder chain pillars and barrier pillars) is practices. It is a shaft mine with nine main and five panel entries but without rooms (pillar recovery constitutes exploitation). Mining is carried out half-advance, half-retreat; that is, the panel entries are driven on the advance and pillars recovered on the retreat. This version of the room and pillar method is called the block system, because pillars are square (they are generally recovered). Details of pillar recovery for the same mine are shown in Fig. 8.3; the extraction technique is called pocket and wing. Other pillar recovery methods are open-ending, slabbing, pocket and stump, and splitting. Additional temporary roof support is usually employed because of the greater risk involved. The pillar line in Fig. 8.3 is staggered in continuous mining, it is usually parallel to the rooms or crosscuts, while in conventional mining, it is angled (ideally, 45°) to provide more working places. Recoveries with and without pillar extraction vary as follows (Table 8.1):

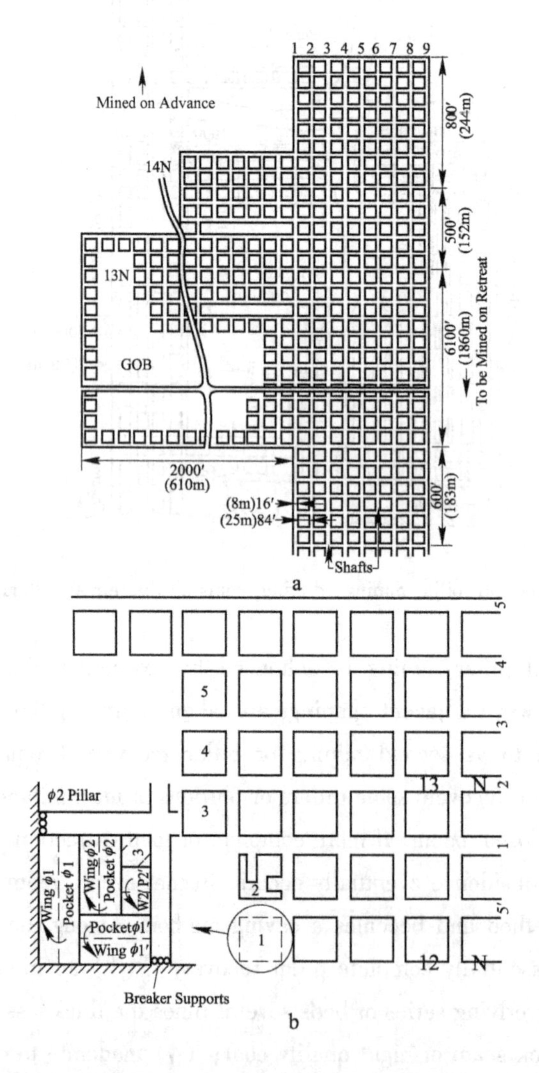

Fig. 8.3 Room and pillar mining, with parital pillar recovery

a—Room and pillar mining, block system; b—details of pillar recovery

Table 8. 1 Recoveries with and without pillar extraction

With or without pillar extraction	Recoveries/%
No pillar extraction, conventional equipment	40~50
No pillar extraction, continuous equipment	50~60
With pillar extraction	70~90

Pillar recovery as a practice is diminishing in coal mining but still far exceeds that in non-coal mines.

Room and pillar mining in the U. S. coal industry has been developed to a high degree of production efficiency. There is likewise a huge disparity in performance among mines, with the following selected as typical for comparable conditions (first mining in 1. 6m seam) with continuous and conventional equipment (Table 8. 2):

Table 8. 2 Production efficiency for comparable conditions (first mining in 1. 6m seam) with continuous and conventional equipment

Items	Conventional equipment	Continuous equipment
Production, t/shift	504	466
Face crew	10	6
Productivity, t/face employee	50	74

It is not unusual for conventional equipment to out-produce continuous, but with a smaller crew, continuous equipment is generally more productive.

Productivity data for the U. S. coal industry have followed a most revealing trend line. Prior to enactment of stringent health and safety legislation in 1969, productivity had risen steadily since World War 2. It then declined rather precipitously and has only recently resumed its upward rise. Today, U. S. underground coal productivity averages 11t per employee-shift, based on processed coal (clean), or about 20t based on run-of mine (raw) coal.

The procedure for the development of flat-bedded deposits, to which room and pillar mining is applicable, was included. A complete mine layout with all development openings was shown in Fig. 8. 2 and Fig. 8. 3.

In sequence, main entries are driven in the deposit from the bottom of the shaft, slope, or outcrop (if a drift mine). Periodically, panel entries and then room entries are turned of at right angles, forming large rectangular blocks or panels in the mineral deposit in which production openings (rooms) are driven and pillars may be extracted. The same development procedure is used for both long-wall and room and pillar mining.

The pattern in which multiple faces are advanced when driving entries or rooms is called the cut sequence. An echelon sequence, popular for advancing entries with a continuous miner in virgin coal, is shown in Fig. 8. 4. The cut sequences are numbered; all cuts numbered 1 are advances before those numbered 2, 2 before 3, and so forth. The law limits each cut to the length of supported roof, under which the machine operators work, which is approximately equal to the length of the machine (about 6m), unless a variance is granted (usually to 12m) with conventional equipment, the advance averages 3m and is limited by the length of the cutter bar.

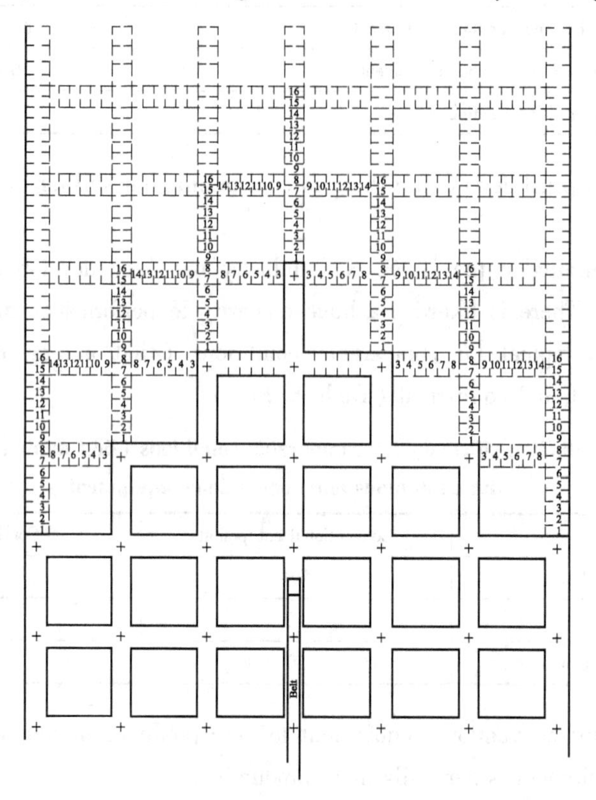

Fig. 8.4 Cut sequence of continuous miner in seven-entry room and pillar mining. The echelon pattern advances the belt-conveyor entry first

Although it is almost always exceeded, the practical minimum number of parallel openings (entries or rooms) which can be advanced at one time in U. S. coal mining is three if conveyor haulage is employed. The restriction is imposed by the legal requirement that a belt conveyor be isolated on neutral or separate split (circuit) of air and that at least one other opening be provided for fresh air and another for exhaust air.

2. Cycle of operations

2.1 Conventional mining equipment

The production cycle of operations in the room and pillar mining of coal with conventional (cyclic) equipment is modified from the basic cycle by the insertion of cutting to improve breakage:

Production cycle = cut + drill + blast + load + haul

The first three of these operations, involved with coal breakage, comprise face preparation. To these productive unit operations are added certain auxiliary operations, including as a minimum, roof support and ventilation. They are integral to the safe performance of mining by the room and pillar method and consequently are programmed into the cycle of essential operations. Cutting is also performed in the mining of softer nonmetallic minerals (potash, salt, trona, etc.) but omitted in hard-rock mining (iron, lead, stone, zinc, etc.).

Because of the highly cyclical, complex nature of room and pillar mining with conventional equipment, it lends itself to the application of mathematical optimization techniques. Computer programs using simulation, a systematic means of evaluating performance and operational factors, are available and ideally suited to optimize the cycle of operations in conventional mining.

Equipment employed in conventional mining and nearly all of it electric powered, consists of the following:

Cutting: short-wall (low-capacity, pulled by cable, undercutting only), universal (high-capacity, rubber-tires, cuts and shears) for coal; omitted in hard rock.

Drilling: mobile auger (rotary drag-bit) drill rig (coal); percussion drill rig (hard rock)-drilling and cutting patterns in coal shown in Fig. 8. 5.

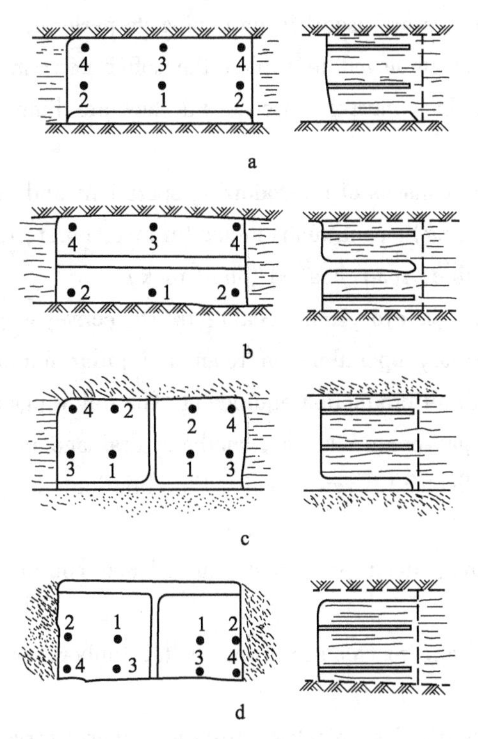

Fig. 8. 5 drilling and cutting patterns for coal breakage. Holes are numbered in order of firing
a—Undercutting; b—Center cutting; c—Undercutting and center shearing; d—Top cutting and center shearing

Blasting (shooting): airdox (compressed air), cardox (compressed carbon dioxide); permissible nitroglycerin and ammonium nitrate explosives (coal); ANFO and slurry explosives (hard rock); loading by machine (bulk) or hand (cartridge), firing by electricity.

Loading: mobile gathering-arm loader; LHD, shovel loader, front-end loader, slusher (hard rock).

Haulage: rubber-tired shuttle car (electric or diesel), belt conveyor, rail; truck (hard rock).

A unique feature of room and pillar mining, pioneered in the U. S. coal industry between the two world wars, is the now-total use of trackless equipment for all face operations. The practice has spread to hard-rock mining, especially with the stope and pillar method.

2.2 Continuous mining equipment

In continuous mining of coal, the basic cycle of operations is simplified as follows:

Production cycle = mine + haul

As in conventional mining, the production cycle is augmented by auxiliary operations of roof control, ventilation, and cleanup. A continuous miner breaks and loads the coal mechanically and simultaneously. It is not, however, as its name implies, truly a "continuously" operating machine; studies have revealed its average productive time use to be only about 20%. Maintenance, moving, and scheduling problems account for much of the inefficiency. While considerably more efficient than the conventional cycle, continuous mining, too, benefits from optimization. Simulation analyses are useful in improving its performance.

Continuous miners are also employed to exploit the softer non-metallics; but in hard-rock mining, tunnel-boring machines must be used, and they are limited to development work in medium-hard rock.

The basic production cycle consists of the following operations and equipment.

Mining (breakage and loading): continuous miner (ripper, borer, milling head, auger, drum) (coal); tunnel-boring machine, road-headed (hard rock).

Haulage: shuttle car, belt conveyor, rail (coal); truck, conveyor (hard rock).

Auxiliary operations. Auxiliary operations in room and pillar mining are very similar for all mineral commodities and with all production equipment. The followings are generally incorporated:

(1) Health and safety: gas control (e. g., methane drainage in coal), dust control (rock dust, water spray, dust collector), ventilation (line brattice, fan and vent tubing), noise abatement.

(2) Environmental control: flood protection, subsidence control, remote monitor to sense atmospheric contamination.

(3) Ground control: scaling, roof control (roof bolts, timber, arch, crib, hydraulic jack), controlled caving.

(4) Power supply and distribution: electric substation, diesel service station.

(5) Water and flood control: pump station, drainage (ditch, pipeline, sump).

(6) Clean up and waste disposal: scoop tram, storage, hoisting, dumping.

(7) Material supply: storage, delivery of supplies.

(8) Maintenance and repair: shop facilities.

(9) Lighting: stationary flood, equipment-mounted fluorescent.

(10) Communications: radio, telephone.

(11) Construction: haulage routes.

(12) Personnel transport: cage, trip, car.

As already mentioned, those considered essential to safety, such as ventilation and roof support, become an integral part of the production cycle in coal mining.

Conditions: The natural and geologic conditions of the mineral deposit which are optimal or well suited for room and pillar mining are the following:

(1) Ore strength: weak to moderate.

(2) Rock strength: moderate to strong.

(3) Deposit shape: tabular.

(4) Deposit dip: low (<15 degree), prefer flat.

(5) Deposit size: large area extent, not thick (<4. 5m) or bench with stope and pillar mining.

(6) Ore grade: moderate.

(7) Ore uniformity: fairly uniform.

(8) Depth: shallow to moderate (<450m for coal and <600m for non-coal, although potash mined at 900m).

Features: We continue to summarize method features as advantages and disadvantages.

Advantages: Moderately high productivity (including both room and pillar and long-wall, based on total work force, U. S. average for clean coal 14t, or 13t per employee-shift; in room and pillar, for raw coal, face productivity 30~80t, or 27~72t, per employee-shift).

Moderate mining cost (relative cost 30%).

Moderately high production rate.

Fair to good recovery with pillar extraction (70%~90%), low to highdilution (0~40%).

Suitable for total mechanization, not labor-intensive.

Concentrated operations (although multiple working places required for cyclic equipment).

Versatile for variety of roof conditions.

Superior ventilation with multiple openings and ifunidirectional (mining and ventilation methods compliment one another).

Disadvantages: Caving and subsidence occur with substantial pillar extraction.

Method fairly inflexible and rigid in layout, not selective without waste disposal.

poor recovery (40%~60%) without pillar extraction, fair (60%~80%) with extraction.

Ground stress and support loads increase with depth.

Fairly high capital investment associated with mechanization.

Extensive development required for coal deposit because of multiple openings.

Potential health and safety hazards underground, especially in coal mines.

Vocabulary

orthogonally [ɔːˈθɒgənl] *adj.* 正交的

tabular [ˈtæbjulə] *adj.* 平板（状）的

nonmetallic [ˌnɔnməˈtælik] *adj.* 非金属的

bolter [ˈbəultə]	*n.* 脱党者；筛子；操作筛粉机器的人
seam [siːm]	*n.* 缝，接缝；层
	vt. 接合
	vi. 生裂缝
gassy [ˈgæsi]	*adj.* 充满气体的
minable [ˈmainəbl]	*adj.* 可开采的
crosscut [ˈkrɔskʌt]	*v.* 横越；横切
	n. 切割横巷
	adj. 横切的
drive [draiv]	*vt.* 掘进
exploitation [ˌeksplɔiˈteiʃən]	*n.* 剥削；开发；利用；采掘
panel [ˈpænl]	*n.* 盘区
drift [drift]	*n.* 巷道
	vt. 漂流；吹积
prevail [priˈveil]	*vi.* 流行；盛行
slab [slæb]	*n.* 厚板
	vt. 制成厚板；使成厚片
stump [stʌmp]	*n.* 树桩；残肢
stagger [ˈstægə]	*vi.* 蹒跚；犹豫
	vt. 使摇晃；使犹豫
	n. 交错
echelon [ˈeʃəlɔn]	*n.* 梯队
	vt. 排成梯队
	adj. 成梯形队伍的
trona [ˈtrəunə]	*n.* 碳酸钠石
auger [ˈɔːgə]	*n.* 锥子，大钻，掘凿钻
nitroglycerin [ˌnaitrəuˈglisərin]	*n.* 硝化甘油
ammonium [əˈməunjəm]	*n.* 氨盐基
cartridge [ˈkɑːtridʒ]	*n.* 弹药筒；容器
LHD (load, haulage and dump)	*n.* 铲运机
shovel [ˈʃʌvl]	*n.* 铲子；挖土机
	vt. 铲起
slusher [sˈlʌʃər]	*n.* 铲泥机；耙斗
shuttle [ˈʃʌtl]	*n.* 往返汽车；梭子
	v.（使）穿梭般来回移动
non-metallics [ˈnɒnmɪtˈælɪks]	*n.* 非金属物质
ripper [ˈripə]	*n.* 粗齿锯操作者；粗齿锯（=ripsaw）
drum [drʌm]	*n.* 鼓，[机械] 鼓轮
ventilation [ˌventɪˈleɪʃən]	*n.* 通风

abatement [əˈbeitmənt]	*n.* 减少
scaling [ˈskeiliŋ]	*n.* 鳞；除铁鳞
crib [krib]	*n.* 饲料槽
	vi. 抄袭；作弊
jack [dʒæk]	*n.* 起重器，千斤顶
ditch [ditʃ]	*n.* 排水沟；壕沟
sump [sʌmp]	*n.* 矿脉底坑；油井；污水坑
scoop [skuːp]	*vt.* 用勺舀；挖空
potash [ˈpɔtæʃ]	*n.* 钾碱；碳酸钾；氢氧化钾
dilution [daiˈluːʃən]	*n.* 冲淡（物），贫化
borax [ˈbɔːræks]	*n.* 硼砂
fluorspar [ˈfluəspɑː]	*n.* 氟石
gilsonite [ˈgilsənait]	*n.* 黑沥青，天然沥青
limestone [ˈlaimstəun]	*n.* 石灰石
gob [gɔb]	*n.* 水兵；（采矿的）填塞材料；（常 pl.）大量
	vi. 吐，吐痰
bituminous [bɪˈtjuːmɪnəs]	*adj.* 沥青的
volatile [ˈvɔlətail]	*adj.* 挥发性的；易变的；轻浮的；爆炸性的；飞行的；能飞的
surge [səːdʒ]	*vi.* （波浪）起伏
	n. 大浪
contingency [kənˈtindʒənsi]	*n.* 偶然性；意外事故；附带发生的事件
width of opening (span)	采场宽度，采场跨度
roof control	顶板控制
Mine Safety and Health Administration (MSHA)	采矿安全与健康管理局
stress distribution	应力分布
adjacent opening	相邻采场
stress super-positioning	应力集中
roof-bolt	顶板锚杆
chain pillar	采场矿柱
barrier pillar	盘区间柱
non-coal mine	非煤矿山
flat-bedded deposit	水平层状矿体
sequence of development	采准顺序
be turned of	转向
virgin coal	原煤
conveyor haulage	输送机运输
noise abatement	噪声消除

exhaust air	污浊空气，尾气
lend itself to	适用于
drag bit	切削钻头
percussion drill rig	冲击式钻机
ammonium nitrate explosive	硝铵炸药
slurry explosive	水胶炸药
ammonium nitrate fuel oil (explosive)	硝酸铵油（爆炸物）
tunnel-boring machine	隧道掘进机，盾构机
methane drainage	沼气排放
non-coal mines	非煤矿山
line brattice	线形板栅
pitch mining	大倾角煤层开采
haulage grade	运输道坡度
multiple seams	多煤层
rock parting	夹石
calorific value	热值

NOTES

[1] In room and pillar mining, openings are driven orthogonally and at regular intervals in a mineral deposit—usually flat-lying (or nearly so), tabular, and relatively thin—forming rectangular or square pillars for natural support. If the deposit and method are very uniform, the appearance of the mine in plan view is not unlike a checkerboard or the intersecting streets and rectangular blocks of a city. Both development openings (entries) and exploitation openings (rooms) closely resemble one another; both are driven parallel and in multiple, and when connected by crosscuts, pillars are formed. Driving several openings at one time increases production and efficiency by providing numerous working places and improves ventilation and transportation as well.

在房柱法中，所开采的矿体通常为水平层状或近似水平层状矿体，矿床中的采场进路间距相等，并且采场进路采用正交布置的方式进行掘进，并形成长方形或正方形矿柱支撑地压。如果矿床和采场进路都规整，则采场进路布置就类似于键盘或城市街道与街区一样。采准巷道（进路）和回采巷道（矿房）高度类似，两者都是多进路平行掘进，当切割横巷联通采准巷道与回采巷道时，采场中就会形成矿柱。同时掘进多条进路就提供的多个作业工作面，可以增加采场生产能力和效率，同时改善采场通风条件和运输条件。

[2] In design practice, the selection of pillar size (or ration of pillar-to-opening widths) and the nature and amount of support are carefully coordinated. When an opening is created, the unit weight of overburden above it is transferred by an arching action to the sides of the opening or adjacent pillars. Because high stresses result, it may be necessary to design the width of the pillar

to avoid stress super-positioning, especially in the case of long-life entries. To do so requires a pillar width equal to at least three times the opening width. For example, if entries are driven 6m in width, their center spacing must be 24m to provide a pillar width of 18m, a ratio of 3 : 1. With such a design, a square roof-bolt pattern on 1.2 ~ 1.5m centers would probably be likely. In exploitation, wider rooms (9m) would probably be driven on closer centers (say, 12 ~ 18m) because of their more temporary nature.

在实际设计过程中，矿柱尺寸大小（或矿柱宽高比）以及矿柱性能和数量需要共同考虑。当采场形成时，采场上覆岩层的重力就以拱形方式转移到了采场周围的矿柱和侧墙之中。因为这些应力比较高，因此，矿柱宽度的设计必须使矿柱内不会出现应力过度集中，特别是在永久巷道内不能出现应力过度集中现象。因此，矿柱宽度至少要等于三倍矿房宽度。例如，如果开采进路宽度为6m，则进路间距必须等于24m才能使矿柱宽度达到18m，此时矿柱与矿房宽度之比为3：1。这种设计中，顶板锚杆为正方形布置，锚杆间距可以是1.2~1.5m。在采矿过程中，由于采场作业时间短，因此掘进的矿房宽度可以增加（如9m宽），而矿房之间的距离缩小（如12~18m）。

Unit 9 Stope and Pillar Mining

全面采矿法

1. Introduction

Strikingly similar to but displaying some unique differences from the room and pillar method, stope and pillar is the most widely used of all underground, hard-rock mining methods. Stope and pillar mining is the unsupported method in which openings are driven horizontally in a mineral deposit in regular or random pattern to form pillars for ground support. It is a large-scale (or intermediate) method, accounting for approximately 50% of U. S. underground non-coal production or nearly 15% of all mineral production. Together with room and pillar, these two similar methods are responsible for over 75% of all underground mining in the U. S..

We will make considerable use of two current sources as we discuss non-coal mining methods, including all the remaining unsupported, supported, and caving (except long-wall mining) classes.

In addition, an older source is mentioned frequently. Generally, a method is classified as stope and pillar rather than room and pillar if it meets two of these three qualifications:

(1) The pillars are irregularly shaped and sized and randomly located. The objective in placement of pillars is to locate them in low-grade ore or waste rather follow a systematic, invariant mining plan, so long as adequate roof support is provided. Because the pillars consist of rock, they are relatively strong, and fewer are required.

(2) The mineral deposit is moderately thick to thick (>6m). If it is sufficiently thick that it cannot be safely or technologically mined in a single pass, then a benching or slabbing technique must be utilized.

(3) The commodity being mined is a mineral other than coal. Although some non-coaldeposit are mined by the room and pillar method, no coal deposits are mined by stope and pillar method.

Rule of thumb for naming method. When in doubt, specify room and pillar if coal; and stope and pillar if non-coal (Exceptions are non-coal mines with a very regular layout of openings and pillars that more properly might be called room and pillar mining).

Other terms that have been applied to stope and pillar mining are open stoping, breast stoping, pillar stoping, but they have generally fallen into disuse.

Two representations of stope and pillar mining are shown in Fig. 9. 1. A less mechanized, less productive version appears in Fig. 9. 1a, applicable to irregular pitching deposits and popular in the period following World War 2. A more modern, more highly mechanized method is also depicted in Fig. 9. 1b, suitable for flatter deposits. Observe that development and production are difficult to distinguish, and in fact, development openings are minimal.

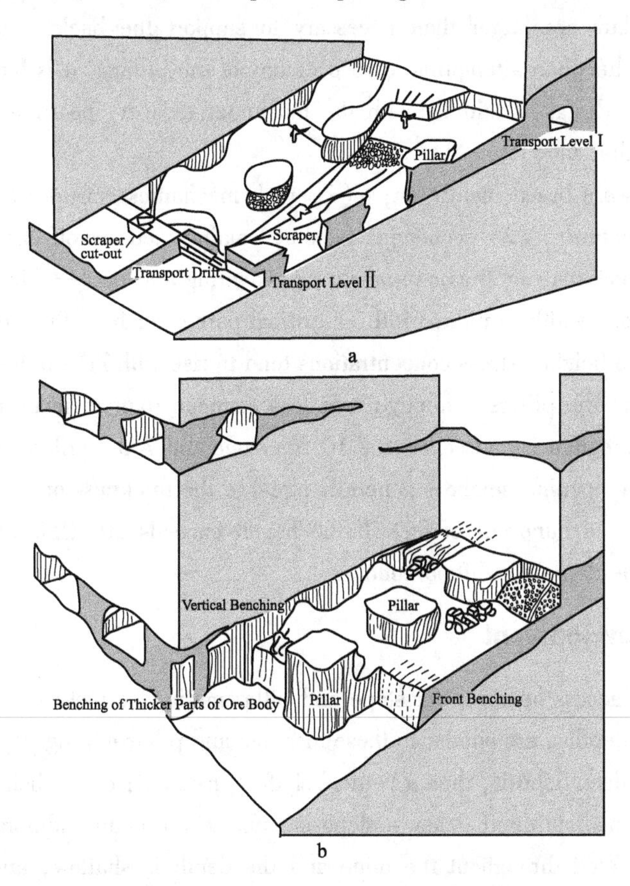

Fig. 9. 1 Stope and pillar mining by benching

a—Intermediate method for pitching deposit using airleg drills, scrapers, and rail haulage;

b—Large-scale method for flat deposit using drill rigs, front-end loaders, and trucks

There are several reasons why the amount of development in stope and pillar mining can generally be sharply reduced over that in room and pillar mining. For one, the strict laws in coal mining regarding multiple openings and ventilation are not applicable to hard-rock mining (unless methane is detected and the mine is classified as gassy), and hence fewer development openings are required and less development must be performed in advance. For another, the deposits to which stope and pillar mining are applied tend to be more irregularly shaped, and development openings cannot always be driven in ore; consequently, fewer are driven (e. g., multiple openings are unusual), and mining on the advance is the rule.

Because the method is less systematic and repetitive, there is less tendency to recover pillars in stope and pillar than room and pillar mining. Total pillar extraction is almost unheard of in a hard-rock mine, for three reasons: (1) the pillars are relatively small and more difficult to recover with safety; (2) the pillars are irregular in size and spacing and do not lend themselves to a systematic recovery operation; and (3) caving to the surface accompanied by subsidence is not usually practiced in non-coal mines. Partial extraction (called pillar robbing) is sometimes practiced, especially if the pillars are larger than necessary to support the back, and some ore can be recovered safely. Rather than attempting to recover any of the pillars, it is likely that a stope and pillar mine would be designed initially with the maximum recovery possible (i. e., the highest possible stope-to-pillar ratio).

Design parameters are based mainly on: (1) rock mechanics considerations, especially of a ground - support nature; (2) economic factors, mainly cut - off grade and recovery; (3) technological concerns, such as equipment maneuvering and grade-ability limits. In addition to the ratio of the stope width to pillar width, a critical parameter from the stress standpoint is the ratio of stope width to height (stress concentrations tend to rise with both ratios). Artificial support is used to supplement the pillars, although it is less common than in room and pillar mining. A typical design with a regular layout consists of 10.5m stope and 7.5m pillars, which yields a pillar recovery of 85%. The opening height is generally equal to the thickness of the deposit, which may range from a few feet to hundreds of feet. If the height exceeds 20~25ft (6~8m), then it is customary to mine the deposit in lifts orbenches .

2. Sequence of development

The choices of main access openings for stope and pillar mining, a method restricted to relatively shallow or moderate depths, are similar to those for room and pillar mining. If skip hoisting is to be used for a relatively deep deposit, then a vertical shaft is installed; or if a belt conveyor is planned for a deposit at intermediate depth, then a slope is sunk. There is an additional choice if rubber-tired haulage is projected throughout the mine and the depth is shallow, and that is a ramp or inclined shaft. Additional openings to the surface, certainly ventilation shafts or raises, are located as required.

Depending on the geometry and attitude of the ore body, secondary openings are constructed on levels connecting the shaft with the production openings. If required by law or good mining practice, parallel drifts and connecting crosscuts may be driven to provide good ventilation. If the mineral deposits are discontinuous and occur on different horizons, then truck haulage and ramps may be selected to provide maximum flexibility.

3. Cycle of operations

3.1 Conventional mining equipment

Because the stope and pillar method is applicable only to non-coal mines, a basic cycle of operations is almost always indicated:

$$\text{Production cycle} = \text{drill} + \text{blast} + \text{load} + \text{haul}$$

Development and production equipment is utilized. If the mineral is sufficiently soft (e. g. salt) that trackless coal-mining equipment can be employed, then the modified cycle which includes cutting is used. If the ore instead is competent, typical of hard-rock mining, then the cycle is modified to include secondary breakage or blasting, especially hazardous in thevertical stoping methods to follow.

The conventional cycle of operations in hard rock consists of the following:

Drilling: pneumatic airleg or pneumatic-or hydraulic-powered, percussion or rotary-percussion drill rig (hard rock); electric-powered rotary drill rig (soft rock).

Blasting: ANFO, gels and slurries, dynamite explosives; charging by hand (cartridge) or by pneumatic loader or pump (bulk), firing electrically or by detonating fuse.

Secondary breakage: drill and blast, impact hammer, drop ball to fragment boulders.

Loading: LHD, front-end loader, shovel, overhead loader, slusher (hard rock); gathering-arm loader (soft rock); LHD equipment reportedly used in 75% of hard-rock mines.

Haulage: truck, rail , LHD, belt conveyor, slusher (hard rock); shuttle car, belt conveyor.

3.2 Continuous mining equipment

The use of continuous mining with the stope and pillar method is infrequent because stope and pillar applications are in hard-rock mines. Tunnels boring machines and other mechanical extractors, however, are used increasingly for development. In softer material, continuous miners are used. The cycle of operations includes the following:

Mining (breaking and loading): tunnel-boring machine, road-header, raise borer, shaft borer (development in soft to medium-hard rock); continuous miners.

Haulage: belt conveyor, shuttle car, truck, rail.

4. Auxiliary operations

There are no significant differences in the auxiliary operations that are performed with any underground mining methods. The list provided for room and pillar mining is therefore applicable to stope and pillar mining as well. The most important auxiliary functions that are performed are health and safety (dust control, face ventilation with fan and tubing, noise abatement); ground control (scaling roof with boom lift if high opening, auxiliary support by rock bolts, timber, shotcrete, cable, steel sets, and arches); power supply and distribution (compressed air, electric, diesel); and water and flood control (sumps, ditches, pipelines, pump stations). Both health and safety and ground control operations are integrated into the production cycle, but less often than in room and pillar coal mining. For particulars on ventilation, see operations on room and pillar mining.

5. Conditions

(1) Ore strength: moderate to strong.

(2) Rock strength: moderate to strong.

(3) Deposit shape: tabular, lens.

(4) Deposit dip: preferably flat, low to moderate (<30°).

(5) Deposit size: any, preferably large area extent, moderate thickness or bench if high (maximum<90m).

(6) Ore grade: low to moderate.

(7) Ore uniformity: variable, leave lean ore or waste in pillars.

(8) Depth: shallow to moderate (<900m, in competent rock, although used to 3000ft in very strong rock).

6. Advantages

Moderate to high productivity U. S. average for non-coal mining 30~50t, per employee-shift; maximum 50~70t, per employee-shift.

Moderate mining cost (relative cost: 30%).

Moderate to high production rate.

High degree of flexibility; method easily modified; can operate on multiple levels simultaneously.

Lends itself readily to mechanization; suitable for large, mobile equipment.

Not labor-intensive; extensive skills not required, similar to surface mining jobs.

Selective method, permitting waste or lean ore to be left in place; suitable for variable deposits.

Early development not extensive.

Multiple working places can be operated.

Fair to good recovery without pillar extraction (range 60%~80%, average 75%), low dilution (10%~20%).

7. Disadvantages

Ground control requires continual maintenance of back, if rock and ore not competent; high back difficult to scale and support; ground stress on openings and pillars increases with depth.

Large capital expenditure required for extensive mechanization.

Some ore lost in pillars.

Difficult to provide good ventilation for dilution of contaminants because air velocities low in large openings.

8. Applications

Stope and pillar mining finds wide application in the exploitation of nonmetallic and metallic; as stated, it is the most popular of the underground non-coal methods. It is used in the mining of copper, iron, lead, limestone and marble, uranium and zinc.

9. Variations

Thick deposits pose somewhat of a challenge in stope and pillar mining, requiring the adoption of benching or slabbing when the opening height reached 20~25 feet. Fig. 9.2 compares the two methods. After breast faces have been advanced a sufficient distance through the deposit at the maximum minable height, then benching or slabbing commences to open up the stopes to their full height. The preferred benching method is shown in Fig. 9.2 (center and bottom), in which the original face is driven to intersect with the intended back (roof) of the stope; this permits benching to recover ore to the lower limit of the deposit. Drill-holes may be drilled horizontally (Figure 6, bottom) or vertically (center). Vertical holes resemble the pattern in surface mining, permitting the extraction of very high benches on the second pass.

Pitch mining is carried out if the deposit dip exceeds the gradeability of the mobile equipment in use. Generally, this occurs if the pitch is >15 degree and is effective to a maximum limit of 30°. Above that inclination, another mining method should preferably be used, although Hamrin proposes a method (step mining), which may extend the range of stope and pillar mining to 45°. The angle of repose of broken ore in the stope (about 40°~50°) is the usual lower limit for the application of vertical stoping methods, and stope and pillar is no longer competitive. A vertical method similar to stope and pillar mining in the supported class is stull stoping.

Vocabulary

pitching [ˈpɪtʃɪŋ] *n.* 俯仰
adj. 倾斜的；陡的
methane [ˈmeθein] *n.* 沼气；甲烷
maneuvering [məˈnuːvərɪŋ] *v.* 移动，用策略；操纵
standpoint [ˈstændpɔint] *n.* 立场，观点，看法
supplement [ˈsʌplimənt] *n.* 附刊；补充；追加
vt. 补充；增补
skip [skip] *vi.* 跳跃
vt. 跳过
n. 箕斗
slope [sləup] *n.* 坡；斜面；倾斜；坡度；斜井
vi. 倾斜；有坡度；走动
vt. 使倾斜；使有坡度
ramp [ræmp] *n.* 斜坡道
raise [reiz] *vt.* 饲养；增加；提高
n. 人行井，增加
drift [drift] *n.* 巷道
vi. 漂流
vt. 漂流
pneumatic [njuːˈmætik] *adj.* 空气的；气体的［物理］气体力学的
percussion [pəːˈkʌʃən] *n.* 撞击；震动；碰击
gels [dʒelz] *n.* 凝胶剂
slurries [ˈsləːriz] *n.* 泥浆，浆状炸药

detonate [ˈdetəuneit]	*vt*. 使爆裂；使爆炸 *vi*. 爆炸，爆发，触发
fuse [fjuːz]	*n*. 保险丝，熔丝 *v*. 熔合；结合
boulder [ˈbəuldə]	*n*. 大的鹅卵石，大块
shotcrete [ˈʃɒtkriːt]	*n*. 喷浆混凝土（喷射混凝土）
cable [ˈkeɪbl]	*n*. 锚索支护
arch [ɑːtʃ]	*n*. 拱支架
ditch [ditʃ]	*n*. 排水沟；掘沟
sump [sʌmp]	*n*. 矿脉底坑；污水坑
lens [lenz]	*n*. 透镜；透镜状矿体
pose [pəuz]	*n*. 姿势 *vt*. 使摆好姿势；使困难
commence [kəˈmens]	*vi*. 开始
grade-ability limit	爬坡能力
stress concentration	应力集中
stope and pillar mining	全面法
room and pillar	房柱法
low-grade ore	低品位矿石，贫矿
non-coal deposit	非煤矿体，非煤矿山（即指金属矿体，非金属矿体）
no coal deposit	没有煤矿体，没有煤矿
rule of thumb	经验法则；根据经验
vertical shaft	竖井
inclined shaft	斜井
ventilation shaft	通风井
connecting crosscut	连接横巷
secondary breakage or blasting	二次破碎或二次爆破
raise borer	天井钻机
road-header	开路机
shaft borer	竖井钻机
scaling roof	撬顶，撬毛石
rock bolt	岩石锚杆
compressed air	压缩空气
pump station	水泵房
deposit dip	矿体倾角
ore grade	矿石品位
lean ore	贫矿
low dilution	贫化率低

gradeability of the mobile equipment 移动设备的爬坡能力
pitch mining 大角度采矿法

NOTES

[1] Strikingly similar to but displaying some unique differences from the room and pillar method, stope and pillar is the most widely used of all underground, hard-rock mining methods. Stope and pillar mining is the unsupported method in which openings are driven horizontally in a mineral deposit in regular or random pattern to form pillars for ground support. It is a large-scale (or intermediate) method, accounting for approximately 50% of U. S. underground non-coal production or nearly 15% of all mineral production. Together with room and pillar, these two similar methods are responsible for over 75% of all underground mining in the U. S.

全面采矿法与房柱法高度相似，但是全面法又具有自身的特点，全面法广泛用于地下硬岩矿床的开采。全面法属于空场法，全面法是在矿体里面水平掘进回采空间，并留下规则矿柱或留下任意矿柱来支撑顶板。全面法是一种大规模（或中等规模）的采矿方法，该采矿方法所采出的矿石产量占美国地下非煤矿石产量的15%。采用全面法与房柱法所采出的矿石量之和占美国地下开采矿石总量的75%。

[2] Generally, a method is classified as stope and pillar rather than room and pillar if it meets two of these three qualifications:

(1) The pillars are irregularly shaped and sized and randomly located. The objective in placement of pillars is to locate them in low-grade ore or waste rather follow a systematic, invariant mining plan, so long as adequate roof support is provided. Because the pillars consist of rock, they are relatively strong, and fewer are required.

(2) The mineral deposit is moderately thick to thick (>6m). If it is sufficiently thick that it cannot be safely or technologically mined in a single pass, then a benching or slabbing technique must be utilized.

(3) The commodity being mined is a mineral other than coal. Although some non-coal deposit are mined by the room and pillar method, no coal deposits are mined by stope and pillar method.

通常，当某种采矿满足下面三个条件中的两个条件，就可以定义为是全面法：

(1) 留下的矿柱形状不规则，大小不一致，矿柱位置不固定。采场内应该选择矿石品位低的矿体留作矿柱，或在采场内把岩石留下作为矿柱。而房柱法采场内留下的矿柱是规则不变的，只要所留矿柱能足以支撑顶板压力。在全面法中，留下岩石矿柱，因为其强度相对较高，因此，采场内的矿柱数量也少。

(2) 矿体厚度为中厚至厚（大于6m）。如果矿体厚度足够大，一次性回采全部厚度的矿体就不安全，技术上也做不到，因此，需要采取台阶式开采或分层开采。

(3) 所开采的矿体为非煤矿体。尽管某些非煤矿体可以采用房柱法开采，但是没有一个煤矿采用全面法开采。

[3] Rule of thumb for naming method. When in doubt, specify room and pillar if coal and stope and pillar if non-coal.

根据经验可以确定采矿法。通常在煤炭开采时就称为房柱法，在非煤矿开采时就称为全面法。

[4] There are several reasons why the amount of development in stope and pillar mining can generally be sharply reduced over that in room and pillar mining. For one, the strict laws in coal mining regarding multiple openings and ventilation are not applicable to hard-rock mining (unless methane is detected and the mine is classified as gassy), and hence fewer development openings are required and less development must be performed in advance. For another, the deposits to which stope and pillar mining are applied tend to be more irregularly shaped, and development openings cannot always be driven in ore; consequently, fewer are driven (e. g., multiple openings are unusual), and mining on the advance is the rule.

全面法开采的采准工作比房柱法要大幅减少，可以从以下几方面说明其原因。首先，在煤炭开采中对多作业面及通风有严格的法律规定，而在硬岩开采中不需要执行这些法律规定（除非这些硬岩矿体也是瓦斯矿体），因此在全面法开采中，需要的采准作业工作面就少，对应的采准巷道就少。其次，多数全面法开采的矿床的形状不规整，采准巷道不会总在矿体里面掘进，因此，掘进的巷道数就少（即多工作面同时掘进的现象少见），在采场里向前掘进就算采矿工作。

Unit 10 Shrinkage stoping

留矿采矿法

1. Introduction

These are classified as a method using artificial support but the support is in fact broken ore left in the stope. After a rock surface is exposed it gradually weathers and relaxes, and is subject to induced stresses from mining operation. If an open stope is left and is a medium strength rock, the sides will spall and slab off, but if the stope is filled with broken ore then the fractured rock is restrained from falling. However shrinkage stoping cannot be used in weak rock because the sides of the stope would squeeze together and trap the broken ore so that it cannot flow.

Shrinkage stoping relies on the fact that when a solid rock is broken by blasting then the broken fragments occupy a larger volume. This expansion is sometimes known as the swell factor. It can be around 1.3 to 1.5 (i. e. a volume increase of 30 to 50 percent) dependent on the amount of fragmentation. If it were not for this swell there would be no output during the stoping period.

After a shrinkage stope has mined out to the top then the broken ore is drawn off leaving the stope empty. The backs under the crown pillar may be rock-bolted to increase their stability and avoid premature collapse when drawing off ores. The Zinc Corporation install water sprays on the backs before the stope is drawn off. The ore is saturated from above to help suppress dust at the draw points.

An alternative method of abandoning a stope is to blast the crown pillar and cave it along with the stope. If intervening pillars are ring blasted then the whole level can be block caved. This may be a way of starting a block cave, which is described later.

A list of the relative merits of shrinkage stoping is given below. Although the disadvantages apparently outnumber the advantages no economic assessment has been attempted. All the relative factors must be assessed when this method is considered.

Shrinkage stoping is the so-called vertical stoping methods. The stope is carried on essentially in a vertical or near-vertical plane at an angle greater than the angle of repose of the broken ore. Shrinkage stoping is an overhand method in which the ore is mined in horizontal slices and remains in the stope as temporary support to the walls and to provide a working platform for the miners. Since the ore swells in breakage, 30%~40% of the broken ore in the stope must be drawn off ("shrunk") during mining to provide sufficient working space. The remainder of the broken ore is recovered when the stope reaches its upper limit. This hold-back in production of

60% ~ 70% of the ore represents a significant tie-up of capital, the cost of which must be charged against the method. On the other hand, it affords storage capacity and the opportunity for blending.

Because of its simplicity and original small operating scale, shrinkage stoping was formerly a very popular method of non-coal mining. Rising costs, the scarcity of skilled labor, and the trend toward mechanization have largely displaced shrinkage stoping. Its use today extends to no more than 1% of U. S. mineral production, where it finds application as a moderate or even small-scale method.

Some method classifications place shrinkage stoping in the supported class because broken ore is left in the stope to provide ground support. This is a temporary measure, however, and like pillars, makes use of the mineral itself. Few shrinkage stopes require the provision of any artificial support other than occasional rock bolts or timber stulls.

The key design parameters in shrinkage stoping are the dimensions of the stope, largely governed by the shape and size of thc dcposit. In a relatively narrow ore body, the stopes are placed longitudinally; in a large or wide ore body, the stopes are located transversely. Stope widths vary from 1 ~ 30m, lengths from 45 ~ 90m, and heights from 60 ~ 90m. Although rock mechanics is a consideration in design, the stope opening itself is relatively small and not excessively stressed, and therefore the major concern is to maintain a managable-sized stope that ensures a smooth flow of ore by gravity and effective draw control.

We consider shrinkage stopingan unsupported methods because the stope opening itself generally has no pillars and very little artificial support. If the ore body is large, and several stopes adjoin one another, then it is customary to leave pillars between stopes for ground control. Rib pillars are left at the ends (sides) of the stope, a sill pillar at the bottom, and a crown pillar at the top. Sometimes these pillars are recovered in subsequent mining, occasionally backfill is placed in the stopes after the ore is drawn down as a permanent means of ground control.

2. Sequence of development

The nature of all vertical stoping methods is such that production operations are carried on over a considerable vertical distance. Consequently, several levels are required, then main or haulage production levels being spaced generally 60 ~ 180m apart. If the stope height is less than the level interval, then sublevels may be constructed, connected by ore-pass.

On each main level, a haulage drift is driven parallel to the strike of the ore body. It is connected to the shaft (usually vertical because of the vertical distance involved) by a haulage crosscut. If the stopes are transverse or draw-points offset, then a haulage lateral or loading cross-cuts to the draw-points are constructed. To avoid multiple access routes into each stope, but mainly to ensure through ventilation, raise man-ways are driven between levels, preferably at the ends of each stope.

The two main tasks of vertical-stope preparation are to construct a means of drawing ore in which muck flows by gravity to the bottom of the stope, and undercut (horizontally) at the sill level or to slot (vertically) the stope, providing an opening into which the ore mainly breaks and subsequently flows. Shrinkage stoping requires undercutting.

To construct the draw system and undercut, finger raises are driven at the desired spacing to connect the haulage level with the sill sublevel. The tops of the finger raises are connected by a small drift which runs the length of the stope; from it, crosscuts are driven above the raises to the walls of the stope. To form bells, slabbing of the finger raises begins at the top, diminishing toward the bottom of each raise, creating funnel-like openings. To form the undercut and provide working space for stoping (a desirable height of 1.8m), the pillars formed by the drift and crosscuts are slabbed off, the broken ore falling into the just-formed bells.

Three draw system are in use today. The oldest, a small-scale method, consists of bells and finger rises terminating in chutes through which haulage conveyances (usually mine cars drawn by a locomotive) are loaded directly (Fig. 10.1). Simple in concept, the gravity and chute system replaced hand loading, but blockages are frequent, and it is no match for draw-point loading. A second system utilizes rope-drawn scrapers, installed in slusher (scram) drifts arrange longitudinally under the stope (Fig. 10.2). The modern and preferred system of draw-points likewise eliminates the bottleneck of chutes and expanding on mechanization, uses loading machines operating at draw-points in cross-cuts under the stope (Fig. 10.3). It is capable of attaining moderate production rates.

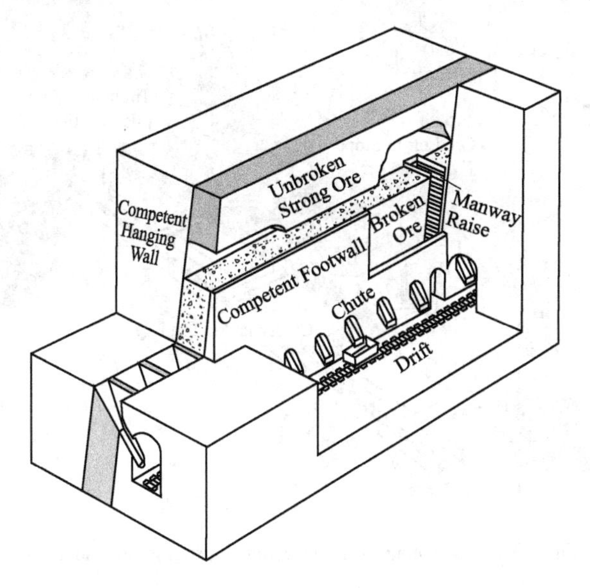

Fig. 10.1 Shrinkage stoping using gravity draw and chutes to load cars

In all draw systems, the spacing of chutes, slusher pockets or draw-points ranges from 5~15m. close spacing requires more development work but minimizes the tonnage of ore stranded on the sill after the stope is emptied and maintains a more even work platform of broken much in the stope. Wide spacing economizes on development at the expense of even draw control and ore loss on the sill (it may be acceptable if a lower stope recovers the sill pillar and stranded much).

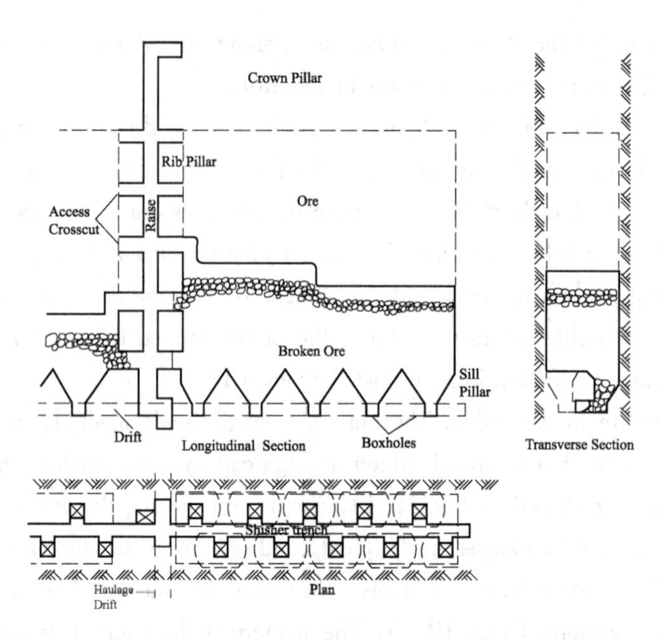

Fig. 10. 2 Shrinkage stoping using a scraper and slusher drift

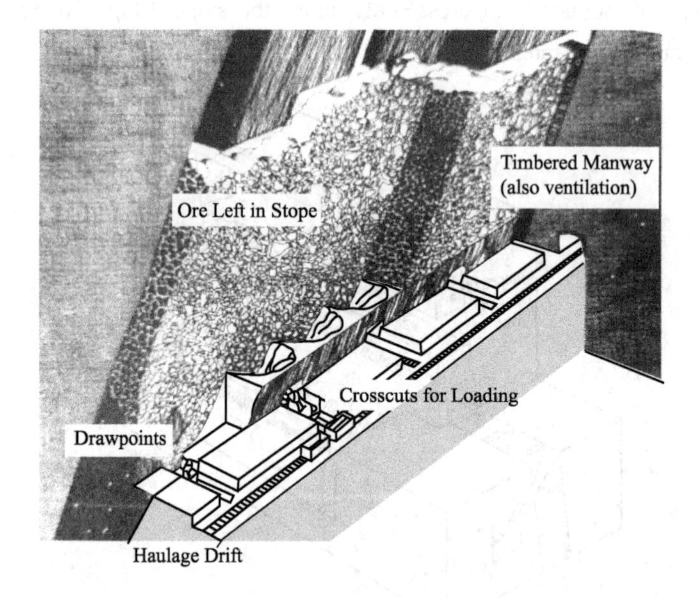

Fig. 10. 3 Shrinkage stoping using drawpoints and loaders

3. Cycle of operations

Working conditions in a shrinkage stope are not ideal because of the uneven working floor of broken muck that periodically is disturbed and lowered by drawing operations in the haulage drift below. It is important to maintain an adequate and safe working space, neither too high nor too low, so that rock-breakage operations within the stope can be conducted properly.

Rock breakage is the principal activity in the stope itself. A bench face is established at the rib pillar and advanced across the face by horizontal drilling with air-legs or vertical drilling with stoppers. Using long-hole drifter drills mounted on columns at the access raise, it is possible to drill out the entire stope length from one setup. After the holes are charged but prior to blasting, drawing of ore from the stope should occur. Before the cycle is repeated, any necessary ground control is carried out, scaling followed by bolting (with wire mesh, if necessary) or timbering (with stulls, less frequently).

Production operations and equipment, all cyclic, in shrinkage stoping follow this basic cycle:

(1) Drilling: pneumatic air-leg, stopper, or drifter percussion drill.

(2) Blasting: ANFO or slurry; charging by hand (cartridge) or by pneumatic loader or pump (bulk), firing electrically or by detonating fuse; blasting by bench rounds for overhand small-scale mining.

(3) Secondary breakage: dynamite bomb, shaped charge, drill and blast to fragment boulders; impact hammer also used.

(4) Loading: gravity flow (through stope); front-end loader, overhand loader, LHD, slusher, chute (under stope).

(5) Haulage: truck, LHD, rail, belt conveyor.

Continuous extraction equipment is rarely used invertical stoping methods because of the nature of the operations and the prevalence of hard rock in most such ore deposits. Its use is mainly limited to development (TBM and raise borer) when the rock hardness permits.

Auxiliary operations. The usual list of auxiliary operations for underground mining pertains——see the complete list for room and pillar mining and the abbreviated list for stope and pillar.

4. Conditions

(1) ore strength: strong (other characteristics: should not pack, oxidize, or be subject to spontaneous combustion).

(2) rock strength: strong to fairly strong.

(3) deposit shape: tabular or lenticular, regular dip and boundaries.

(4) deposit dip: fairly steep (>45°~50° or angle of repose, preferable 60°~90°).

(5) deposit size: narrow to moderate width (1~30m), fairly large extent.

(6) ore grade: fairly high.

(7) ore uniformity: uniform.

(8) depth: shallow to moderate (<750m).

5. Features

5.1 Advantages of shrinkage stoping

(1) Shrinkage stopes in good conditions can be cheaper than cut and fill operations.

(2) Broken ore in the stope can act as a stockpile to even out production operations.

(3) Uphole shrinkage stopingis basically more efficient than cut and fill because it is not a cyclic operation.

(4) There are no ore passes in the stope to wear out or need maintenance or erection.

(5) No ore has to be handled in the stopes, although some scraping may be necessary to level out a working platform for a mobile drill.

5.2 Disadvantages of shrinkage stoping

(1) Wall and roof conditions in the stope must both be good. Ideally the ore body should be vertical but dips to 500 can be coped with.

(2) Stripping a footwall is difficult because sharp ledges must not be formed. These would cause hang-ups; either ore must be left in walls or mullock taken to provide a smooth profile.

(3) Spalling rock from the walls will cause dilution of the ore.

(4) Broken ore may oxidize with possible attendant difficulties in drawing off and in flotation.

(5) There is also a possibility of spontaneous combustion in ores with a high sulphide content.

(6) Closely spaced draw points are needed to give good ore flow. Draw control is difficult and working floor conditions are potentially dangerous. Travelling in the stope is rough.

(7) Large blocks of ore that would normally require secondary blasting may be buried in the stope. These can block a draw point or cause hang-ups.

(8) Shrinkage stopes are slow producer in the mining period. Once the stope is started it is difficult to change to another method.

(9) Because approximately 60 percent of the broken ore remains in the stope until it is completed the interest charges for the money spent on breaking this ore should be charged in the economic analysis.

(10) Shrinkage stoping with transverse pillars needs more expensive development than does a cut and fill stope. About 50 percent of the ore remains as pillars which can present a recovery problem.

(11) Large amounts of sand and development rock are required to fill an empty stope. If the fill is urgently required for stability.

Then there is a big problem.

(12) Manpower requirements of a shrinkage stope can be high because it is difficult to mechanize the drilling operation, and drawing is not independent of the drilling operations while mining. Even at a later stage men may be needed to clear hangups in the stope during the emptying operations.

Vocabulary

spall [spɔ:l] *n.* 碎片；碎石块
vt. 将……弄成碎片
vi. 剥落；裂成碎片

slab [slæb]	*n.* 厚板
	vt. 制成厚板；使成厚片
squeeze [skwiːz]	*vt.* 压榨
	vi. 挤压
	n. 压，拥挤；紧抱
premature [ˌpreməˈtjuə]	*n.* 过早发生的事物；早产儿
	adj. 早熟的
tie-up [ˈtaɪʌp]	*n.* 束缚
	v. 系；约束
blending [ˈblendiŋ]	*n.* 混合（物），配矿
stull [stʌl]	*n.* 横梁；支柱
sublevel [ˈsʌbˌlevəl]	*n.* 分段；中段平坑；顺槽
offset [ˈɔːfset]	*n.* 分支；补偿
undercut [ˈʌndəˈkʌt]	*n.* 拉底
slot [slɔt]	*n.* 拉槽
muck [mʌk]	*n.* 泥土；岩石
crosscut [ˈkrɔskʌt]	*v.* 横越；横切
	n. 切割横巷
	adj. 横锯的；横切的
bell [bel]	*n.* 漏斗
funnel [ˈfʌnl]	*n.* 漏斗
	vt. 使成漏斗状；使通过漏斗，使穿越
chute [ʃuːt]	*n.* 斜槽
	vt. 以斜槽运送
	vi. 以斜槽下滑
blockage [ˈblɔkidʒ]	*n.* 封锁；妨碍；闭塞；堵塞物
scraper [ˈskreipə]	*n.* 耙子
slusher [sˈlʌʃər]	*n.* 铲泥机；耙斗
scram [skræm]	*vi.* 走开
	n. 紧急关闭快速，停止
longitudinally [ˌlɒndʒɪˈtjuːdɪnli]	*adv.* 经度地，纵观地，纵行地
strand [strænd]	*vt.* 使搁浅
	vi. 搁浅
stopper [ˈstɔpə]	*n.* 塞子，阻挡者，上向凿岩机
	vt. 给……配上塞子
air-leg [ˈeərlˈeg]	*n.* 气腿式凿岩机
scaling [ˈskeiliŋ]	*n.* 撬毛石，除铁鳞
wire mesh	*n.* 金属网
timbering [ˈtimbəriŋ]	*n.* 支架，木支架

dynamite [ˈdainəmait]	*n.* 炸药
	vt. 用炸药炸毁；爆破；毁坏
boulder [ˈbəuldə]	*n.* 大的鹅卵石
pack [pæk]	*n.* 包裹
	vt. 捆扎；塞满
lenticular [lenˈtikjulə]	*adj.* 透镜状的
exploration [ˌeksplɔːˈreiʃən]	*n.* 探险；研究；探查
development [diˈveləpmənt]	*n.* 开拓
exploitation [ˌeksplɔiˈteiʃən]	*n.* 采矿；采掘
dike [daik]	*n.* 堤；排水道；岩脉
	vt. 筑堤保护，用堤围绕
	vi. 筑堤
porphyry [ˈpɔːfiri]	*n.* 斑岩
rhyolite [ˈraiəlait]	*n.* 流纹岩
magnetite [ˈmægnitait]	*n.* 磁铁矿
hematite [ˈhemətait]	*n.* 赤铁矿，铁的主要原矿
pellet [ˈpelit]	*n.* 球团矿
backfilling [ˈbækfɪlɪŋ]	*n.* 充填
crude ore	原矿
shaped charge	聚能装药
induced stress	诱发应力
premature collapse	过早冒落
swell factor	碎涨系数
crown pillar	顶柱
drawing off ore	出矿
intervening pillar	中间支柱，间柱
ring blasted	环形孔爆破
overhand method	上向采矿法
hold back	阻止，停滞
tie-up of capital	资金积压
be charged against	被控告
rib pillar	间柱
sill pillar	底柱
crown pillar	顶柱
haulage production level	运输生产水平，中段，阶段
ore-pass	溜矿井
the strike of the ore body	矿体走向
haulage crosscut	石门
loading cross-cut	装载横巷

haulage lateral	穿脉运输巷
raise man-way	人行天井
draw system	出矿系统
finger raise	手指状天井，指状天井
no match	不匹配
rope-drawn scraper	电耙
arrange longitudinally	沿走向布置
slusher pocket	斗穿
at the expense of	以……为代价
access raise	人行天井
gravity flow	重力放矿
overhand loader	上举式装载机
be subject to	易于……
spontaneous combustion	自燃性
deposit dip	矿体倾角
rhyolite porphyry	流纹斑岩
pneumatic cartridge loader	气动装药器
slurry pump	浆状炸药装药器
belt conveyor	皮带输送机
skip pocket	箕斗装载硐室

NOTES

[1] These are classified as a method using artificial support but the support is in fact broken ore left in the stope. After a rock surface is exposed it gradually weathers and relaxes, and is subject to induced stresses from mining operation. If an open stope is left and is a medium strengthrock, the sides will spall and slab off, but if the stope is filled with broken ore then the fractured rock is restrained from falling. However shrinkage stoping cannot be used in weak rock because the sides of the stope would squeeze together and trap the broken ore so that it cannot flow.

留矿法是一种使用人工支护的采矿方法，但是这种人工支护是指用采场破碎矿石留在采场内进行的支护。当岩石表面暴露在空气中，岩石就会逐渐风化剥蚀，并受到采矿的扰动应力影响。如果采场暴露，并且其岩石属中等强度，则采场的侧墙将产生剥离和片帮，但是，如果采场内充满破碎矿石，则采场内破裂岩石会受到破碎矿石的支撑，从而减少采场内的片帮剥离现象。可是，留矿法不能用于破碎岩石的开采，主要是矿体两侧的岩石破碎容易向采场内挤压变形，采场空间变窄，使采场内破碎矿石的放矿更加困难。

[2] Shrinkage stoping relies on the fact that when a solid rock is broken by blasting then the broken fragments occupy a larger volume. This expansion is sometimes known as the swell factor. It

can be around 1. 3 to 1. 5 (i. e. a volume increase of 30 to 50 percent) dependent on the amount of fragmentation. If it were not for this swell there would be no output during the stoping period.

当固体岩石爆破成碎块后，岩石破碎后的体积增大，这一特性是留矿法应用的基本条件。岩石破碎后体积增加的性能也叫碎胀性。其碎胀系数约为 1. 3~1. 5（即岩石破碎后的体积增加了 30%~50%），碎胀系数与破碎程度有关。如果没有碎胀，那么采场内就不能出矿。

[3] After a shrinkage stope has mined out to the top then the broken ore is drawn off leaving the stope empty. The backs under the crown pillar may be rock-bolted to increase their stability and avoid premature collapse when drawing off ores. The Zinc Corporation install water sprays on the backs before the stope is drawn off. The ore is saturated from above to help suppress dust at the draw points.

当留矿法采场到达顶板时，则采场内的矿石就可以全部出空。采场顶柱可以采用岩石锚杆进行支护，增加其稳定性，避免顶板在出矿过程中过早冒落。锌矿公司在采场出空之前就在采场顶板配备了喷水设备。在采场内喷洒水，湿润矿石，抑制采场出矿口的粉尘。

[4] Shrinkage stoping is the so-called vertical stoping methods. The stope is carried on essentially in a vertical or near-vertical plane at an angle greater than the angle of repose of the broken ore. Shrinkage stoping is an overhand method in which the ore is mined in horizontal slices and remains in the stope as temporary support to the walls and to provide a working platform for the miners. Since the ore swells in breakage, 30%~40% of the broken ore in the stope must be drawn off ("shrunk") during mining to provide sufficient working space. The remainder of the broken ore is recovered when the stope reaches its upper limit. This hold-back in production of 60%~70% of the ore represents a significant tie-up of capital, the cost of which must be charged against the method. On the other hand, it affords storage capacity and the opportunity for blending.

留矿法是一种所谓的垂直作业空间的采矿法。采场基本为垂直或近似垂直，采场与水平面的夹角大于破碎矿石的自然安息角。留矿法是一种水平分层上向开采的采矿法，留在采场内的矿石作为侧墙的临时支护，并为矿工提供作业平台。由于矿石有碎胀性，采场内爆破下来的矿石需要除掉 30%~40%（减少由于矿石破碎所增加的空间），为矿工提供足够的作业空间。当采场开采到顶部时，留在采场内的矿石就可以全部放出。留在采场内的矿石占采场爆破矿石的 60%~70%，这部分矿石长期滞留在采场内，不能及时放出，造成大量资金积压，这种资金积压带来的成本是这种采矿方法的缺点。但是另一方面，采场内又可以储藏矿石，能为矿山配矿提供条件。

Unit 11 Sublevel Stoping

分段矿房法

1. Introduction

A principal distinction of sublevel stoping is that it is the only patented mining methods (more correctly, a modern version of the method, vertical crater retreat, VCR, is patented). Sublevel stoping is an overhand, vertical stoping method utilizing longhole drilling and blasting carried out from sublevels to break the ore. The ore flows through the stope by gravity in the customary way and is drawn off at the haulage level. Always a popular method, sublevel stoping is enjoying a revival in underground mining, presently having application for about 9% of U. S. non-coal production or 3% of all mineral production. It is classified as a large-scale method. Sublevel stoping, the last unsupported method employs even less temporary support in the stopes than the stope and pillar and shrinkage methods. The reason is that no personnel are exposed in the stope proper; drill and blast crews work in the protective cover of sublevel drifts and crosscuts, while loading and haulage crews labor in the security of haulage drift below.

If support is required in the sublevels, it is readily provided by rock bolts, wire mesh, cables, or shotcrete. Although the stopes are unsupported, pillars are usually left between stopes and occasionally within stopes.

The two common versions of sublevel stoping are shown in Fig. 11.1 and Fig. 11.2. Both employ longhole drilling, the one a ring pattern with small holes (Fig. 11.1) and the other parallel drilling with large holes (Fig. 11.2). In ring drilling, a vertical slot is opened, whereas either a slot or an undercut is used with parallel drilling (the VCR method requires parallel holes and an undercut).

The unique feature of the VCR method is the application of blasting theory in the proper placement of explosives to yield improved fragmentation, reduced stope-wall damage, and increased production.

Ring drilling with blasting into a vertical slot was the original version of sublevel stoping. The drillholes are relatively small (2~3in., or 50~70mm), bored with percussion rock drills mounted on a column and bar (drifter) or fandrill rig and with extension drill steel to a maximun length of 80~100ft (24~30m).

To effect economies in rock breakage, the holes are loaded heavily and a ring (5~10ft, or 1.5~3m, in thickness) detonated simultaneously; the effect is not unlike that in bench blasting.

Hole deviation is a serious problem, however, in which deflections of several feet are not

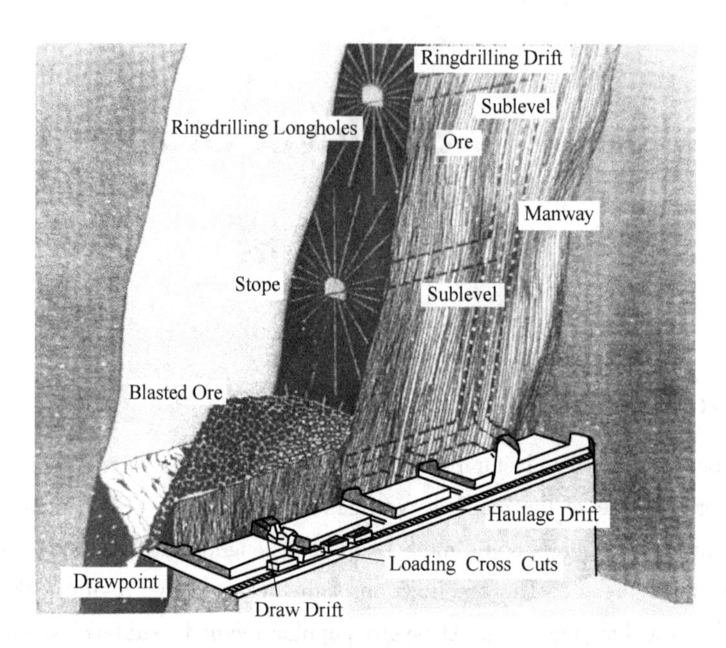

Fig. 11.1 Sublevel stoping (blasthole method) using ring drilling and blasting into a slot

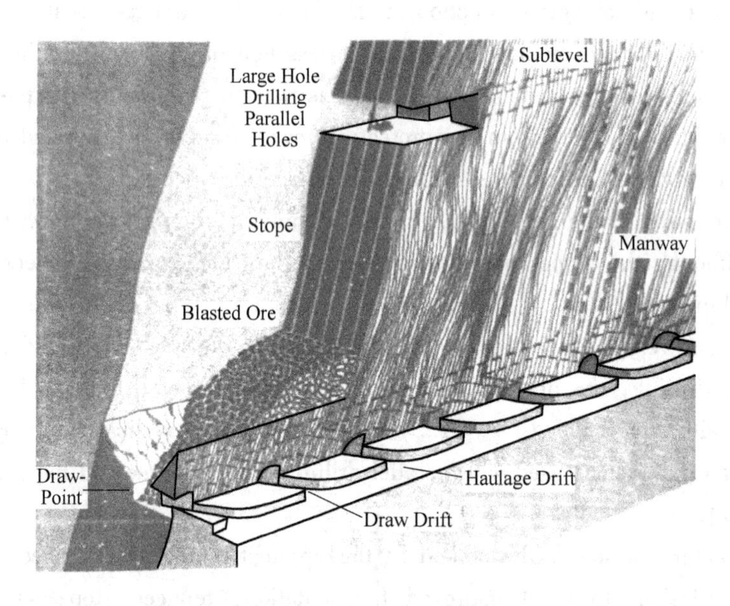

Fig. 11.2 Sublevel stoping using parallel drilling and blasting into a slot

uncommon. The effect in blasting can be disastrous, because ring drilling requires accuracy in hole placement to obtain proper fragmentation.

With the advent of large-diameter (7in. or 200mm) rotary and downhole percussion drills, it became practical to adopt the parallel-drilling method of sublevel stoping (Fig. 11.2). Deviation is no longer a problem with large, parallel holes, which can now be extended to a maximum of 300ft (90m), and sublevel spacing can be increased accordingly.

A further improvement has been realized in the last decade with the advent of the VCR method (Fig. 11.3). Large parallel holes are used as in the parallel-drilling method, but a major innovation has been blasting horizontal slices of ore with near-spherical charges into the undercut. Spherical placement of explosives is the most efficient in terms of fragmentation and powder consumption.

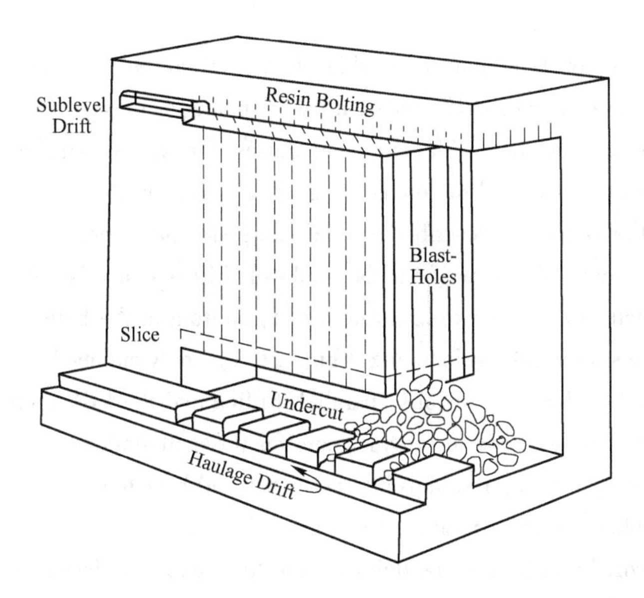

Fig. 11.3 Vertical crater retreat (VCR) version of sublevel stoping. Large parallel holes are loaded with near-spherical explosive charges, and horizontal slices of ore are blasted into the undercut

Holes are charged from the collar after plugging the opposite end; the size of charge is restricted to a length-to-diameter ratio of 6 : 1, which suffices in practice to simulate a spherical charge. All holes in a stope are detonated together. When the broken ore in the stope has been drawn down sufficiently, then the next slice of ore (4.5m in thickness) is blasted.

Because drilling is carried out from a sublevel and is usually complete before blasting commences, unit operations with the VCR method can be conducted with high efficiencies and productivities.

Many of the design parameters in sublevel stoping involve the same factors and dimensions as in shrinkage stoping of any other form of vertical stoping. Stope width varies from 6~30m, maximum length 90m, and maximum sublevel heigth 90m.

Boundary pillars are located similarly to those in shrinkage stoping. Rock mechanics, blasting, and materials-handling considerations are the major concerns in specifying dimensions.

In addition, because of the unique reliance on longhole drilling and blasting, special attention in sublevel stoping must be paid to rock-breakage design; hole diameter and length, burden, explosive selection, powder factor, and so forth. Simulation models and cost analysis are profitably employed.

2. Sequence of Development

Sublevel stoping differs from many other vertical stoping and caving methods in utilizing one or more sublevels between main haulage levels. That distinction is less marked today, however, because other methods have introduced intermediate levels as stope heights have increased.

The general sequence of development in sublevel stoping parallels that in shrinkage stoping and other verical methods. A haulage drift, crosscut, draw points or draw drifts and trench are driven for materials handling, together with interlevel raises for access and ventilation. Either an undercut or end slot is constructed to commence stoping operations.

If an undercut, the sequence described in shrinkage stoping is employed. If a slot, sublevel crosscuts are driven across the stope from the sublevel drift and a raise driven at the boundary. These are then blasted in a manner to excavate a slot, starting at the bottom.

In ring-drilling version of sublevel stoping (Fig. 11.1), only sublevel drifts must be driven for longhole drilling. In the other two versions, parallel drilling and the VCR method, a horizontal slot must be driven across the ore body to provide room for the drill stations.

This is constructed by driving a sublevel crosscut the width of the stope and then advancing it the length of the sublevel drife as mining progresses.

Pillars (sill, crown, and rib) may be delineated at the stope boundaries and left permanently or recovered in a retreat operation, often after backfilling.

3. Cylce of Operation

Like many of the vertical stoping methods, rock breakage and materials handling are carried out in separate sectionsof sublevel stopes.

Drilling and blasting are conducted in the stope proper in sublevel drifts, while the loading and haulage take place underneath the stope in the draw points or draw drifts. Coordination is necessary, of course, but the two major groups of unit operations of the production cycle are carried out largely independently of one another.

The cycle of operations follows the basic production cycle:

(1) Drilling: 1) longhole pneumatic percussion drill (small hole) with coupled steel; 2) downhole pneumatic percussion drill (large-hole) on drill platform of rig; 3) roller-bit rotary drill on drill platform or rig

(2) Blasting: ANFO, slurries, charging by cartridge or bulk pneumatic loader or pump, Secondary breakage: impact hammer, drill and blast.

(3) Loading: gravity flow to draw points; LHD.

(4) Haulage: LHD, truck, rail, belt conveyor.

4. Conditions

(1) ore strength: moderate to strong, may be less competent than for stope and pillar.

(2) rock strength: fairly strong to strong.

(3) deposit shape: tabular or lenticular.

(4) deposit dip: fairly steep (greater than 45°~50°, preferably 60°~90°).

(5) deposit size: fairly thick to moderate width (6~30m), fairly large extent.

(6) ore grade: moderate.

(7) ore uniformity: fairly uniform to uniform.

(8) depth: moderate (less than 1200m), deep (less than 2400m).

5. Features

(1) moderate to high productivity (range 15~30t, or 14~27t, per employee-shift; maximum 30~40t, or 27~36t, per employee-shift).

(2) moderate mining cost (relative cost: 40%).

(3) moderate to high production rate.

(4) lends itself to mechanization; not labor-intensive.

(5) low breakage cost; fairly low handling cost.

(6) little exposure to unsafe conditions; easy to ventilate.

(7) unit operations can be carried on simultaneously.

(8) fair recovery (75%); fair dilution (20%).

6. Disadvantages

(1) fairly expensive; slow and complicated development, high development cost.

(2) inflexible and non selective.

(3) long hole drilling requires careful alignment (less than 2% deviation).

(4) large blasts may cause excessive vibration, air blast and structural damage.

7. Applications

Applications of sublevel stoping, while most common in metal mines, are frequent and varied in hard-rock mines in general. They include copper, copper-iron-zinc-sulfur, copper-lead, gold and nickel mines.

Case study. Copperhill copper-iron-zinc-sulfur mine, Tennessee Chemical Company:

(1) General: only major underground primary producer of sulfur (as sulfuric acid); located in Copper Basin in southern Appalachians; Mining dates to 1847.

(2) Geology: massive sulfide lenses in metamorphosed schists and graywackes; mineralization.

(3) Ore: mine ore bodies extending to at least 900m and up to 120m in thickness; reserves 23000000~63000000t.

(4) Production: 2.2million t/y averaging 0.7% Cu, 20% S and 0.5% Zn.

(5) Mining methods: open pit mining, sublevel stoping.

(6) Equipment (undergoing): drifter jumbos, raise borers, percussion fan-drills, rotary drill, LHDs, trucks, rail.

Variations: The principal variation of sublevel stoping is the widely used Vertical Crater Retreat, (VCR) method (Fig. 11.3), already discussed.

Vocabulary

revival [ri'vaivəl]	*n.* 复活；复兴
undercut [ˌʌndə'kʌt]	*n.* 拉底空间
fragmentation [ˌfrægmən'teiʃən]	*n.* 裂成碎片；破碎
deviation [ˌdiːvi'eiʃən]	*n.* 离题；偏差
deflection [di'flekʃən]	*n.* 歪斜；偏向
advent ['ædvənt]	*n.* 来临；出现，到来
collar ['kɔlə]	*n.* 环管
	vt. 控制；捉住
plug [plʌg]	*vt.* 接上；用塞子塞住
	n. 插头；塞子
boundary ['baundəri]	*n.* 界线；分界；边界
burden ['bəːdn]	*n.* 抵抗线
trench [trentʃ]	*n.* 沟渠
	vi. 掘沟
delineate [di'linieit]	*vt.* 描画；记述
fair [feə]	*adj.* 公平的；应当的
graywackes [greɪ'wækəs]	*n.* 硬砂岩
mineralization [ˌmɪnərəlaɪ'zeɪʃən]	*n.* 矿化作用，成矿作用
commensurate [kə'menʃərit]	*adj.* 等量的；相称的；成比例的
tectonic [tek'tɔnik]	*adj.* 构造的；筑造的；建筑的
reduce [ri'djuːs]	*vt.* 缩减；减少；[化学] 使还原
situ ['sɪtjuː]	*n.* 原地，现场
penetration [ˌpeni'treiʃən]	*n.* 渗透；穿透力；凿岩
inexorable [in'eksərəbl]	*adj.* 无情的；残酷的；坚决的
unrelenting ['ʌnri'lentiŋ]	*adj.* 不宽恕的；冷酷的
elastic [i'læstik]	*adj.* 有弹性的；可伸缩的
homogeneous [ˌhɔmə'dʒiːniəs]	*adj.* 类似的；同类的；同性质的；均一的
isotropic [ˌaisəu'trɔpik]	*adj.* 等方性的，各向同性的
departure [di'paːtʃə]	*n.* 离开；出发；变更；偏差
sublevel stoping	分段空场法
last unsupported method	最新的空场采矿法
vertical crater retreat (VCR)	垂直深孔球状药包后退式采矿法
long hole	深孔
haulage level	出矿水平

sublevel drift	分段巷道
vertical slot	切割槽
percussion rock drill	冲击式凿岩机
mounted on a column	安装在钻架之上
fandrill rig	扇形孔钻机
extension drill steel	接杆
bench blasting	台阶爆破
parallel-drilling	钻凿平行孔
blasting horizontal slices of ore	爆破水平矿层
near-spherical charge	似球状药包装药
spherical placement of explosives	球状药包装药
powder consumption	单位炸药消耗量
plugging the opposite end	堵塞炮孔另一端
boundary pillars	采场周围的矿柱（顶柱，底柱，间柱）
longhole drilling and blasting	深孔凿岩爆破
hole diameter and length	炮孔直径与深度
powder factor	爆破作用指数
draw point	出矿口，出矿点
draw drift	出矿巷道
interlevel raise	中段之间的天井
end slot	端部切割槽
horizontal slot	水平槽，凿岩硐室
retreat operation	退采
coupled steel	接杆
downhole pneumatic percussion drill (large-hole)	潜孔钻机
roller-bit rotary drill	牙轮钻机
massive sulfide lenses	大型硫化透镜矿体
metamorphosed schists	变质片岩
drifter jumbos	平巷掘进机
raise borers	天井钻机
percussion fan-drills	冲击式扇形钻机
rotary drill	回转钻机
sloughing ground	地表崩落
roof fall	冒顶
overlying rock	上覆岩层
rock mechanics	岩石力学
superincumbent load	覆岩层荷载
static rock mechanics	岩石静力学
dynamic rock mechanics	岩石动力学

NOTES

[1] A principal distinction of sublevel stoping is that it is the only patented mining methods (more correctly, a modern version of the method, vertical crater retreat, VCR, is patented). Sublevel stoping is an overhand, vertical stoping method utilizing longhole drilling and blasting carried out from sublevels to break the ore. The ore flows through the stope by gravity in the customary way and is drawn off at the haulage level. Always a popular method, sublevel stoping is enjoying a revival in underground mining, presently having application for about 9% of U. S. non-coal production or 3% of all mineral production. It is classified as a large-scale method. Sublevel stoping, the last unsupported method employs even less temporary support in the stopes than the stope and pillar and shrinkage methods. The reason is that no personnel are exposed in the stope proper; drill and blast crews work in the protective cover of sublevel drifts and crosscuts, while loading and haulage crews labor in the security of haulage drift below.

分段矿房法有其自身特点，该方法是唯一获得专利的采矿方法（该方法是现代发明的采矿方法，即倒漏斗爆破后退式回采的采矿法，简称 VCR 法）。分段矿房法采用上向式开采，采场垂直向上回采，深孔凿岩，从分段水平开始爆破。矿石靠自重在采场内自溜至出矿口，再水平运输出矿。分段矿房法应用广泛，该采矿法正为地下采矿方法提供了广阔的应用前景，该方法在美国的非煤地下矿山开采的矿山量约占 9%，或占所有地下矿石产量的 3%。分段矿房法的生产能力大，它是最新纳入空场法的一种采矿方法。因为在分段矿房法中，作业人员无须进入采场，钻孔与爆破人员都是在支护条件较好的分段巷道及切割横巷内进行作业，装载与运输则在下部运输巷道内进行，安全性好，因此该方法所需要的临时支护比全面法和留矿法少。

[2] If support is required in the sublevels, it is readily provided by rock bolts, wire mesh, cables, or shotcrete. Although the stopes are unsupported, pillars are usually left between stopes and occasionally within stopes.

如果分段巷道内需要支护，则可以采用岩石锚杆、金属网、锚索或喷射混凝土来进行支护。尽管采场内不进行支护，但是在采场之间通常需要留下矿柱，有时在采场内也留下矿柱。

[3] The two common versions of sublevel stoping are shown in Fig. 11. 1 and Fig. 11. 2. Both employ longhole drilling, the one a ring pattern with small holes (Fig. 11. 1) and the other parallel drilling with large holes (Fig. 11. 2). In ring drilling, a vertical slot is opened, whereas either a slot or an undercut is used with parallel drilling (the VCR method requires parallel holes and an undercut).

图 11. 1 和图 11. 2 为分段矿房法的两种主要形式。这两种形式的分段矿房法都采用深孔凿岩。其中一种为小孔径的环形深孔布置（如图 11. 1 所示），另一种为大孔径平行深孔布置（如图 11. 2 所示）。在环形深孔凿岩布置中，采场内需要开掘一个切割槽，而在平行

深孔布置的采场内，即可开掘切割槽，也可以进行拉底（VCR 法需要采用平行深孔布置，进行拉底作业提供初始爆破条件）。

[4] The unique feature of the VCR method is the application of blasting theory in the proper placement of explosives to yield improved fragmentation, reduced stope-wall damage, and increased production.

VCR 法利用了爆破理论的特征，在钻孔内适当位置安设炸药，改善了爆破块度，降低了爆破对采场围岩的破坏作用，提高了采场生产能力。

[5] Ring drilling with blasting into a verticalslot (Fig. 11.1) was the original version of sublevel stoping. The drillholes are relatively small (2 ~ 3in., or 50 ~ 70mm), bored with percussion rock drills mounted on a column and bar (drifter) or fandrill rig and with extension drill steel to a maximun length of 80~100ft (24~30m).

分段矿房法最早采用的是环形炮孔布置，采用垂直切割槽作为补偿空间（如图 11.1 所示）。钻孔直径相对较小（2~3in，或 50~70mm），采用冲击式凿岩机钻孔，凿岩机架设在立柱或环形钻架之上，采用接杆凿岩，最大凿岩深度可达 24~30m。

Unit 12 Supported Methods

充填法

1. Classification of methods

The second category of underground mining methods is referred to as the supported methods. Supported methods are those methods that require some type of backfill to provide substantial amounts of artificial support to maintain stability in the exploitation openings of the mine. Supported methods arc uscd when production openings will not remain standing during their life and when major caving or subsidence cannot be tolerated. In other words, the supported class is employed when the other two categories of methods, unsupported and caving, are not applicable.

Pillars of the original rock mass are the ultimate form of ground control in an underground mine because they are capable of providing near-rigid support. In many horizontal methods, natural pillars are relied on for primary support, supplemented by light artificial supports such as bolts and timber. Vertical methods may use natural pillars to surround the stopes, but pillars in the production areas are impractical because they interfere with production operations. Artificial support in the form of pillars of backfill material is therefore used to control the rock mass and make mine output safe and productive.

In the design of artificial support systems for mining methods, an evaluation, preferably quantitative, of the load-carrying capacity of the natural rock structure is a prerequisite. Rock mechanics tests are performed to evaluate the structural properties of the rock. However to fully understand the capability of the rock to sustain a load, it may be better to determine the rock quality designation (RQD) based on drill core evaluation. The supported class of methods is intended for application to rock ranging in competency from moderate to incompetent.

2. Cut-and-fill stoping

Cut-and-fill stoping is normally used in an overhand fashion. The ore is extracted in horizontal slices and replaced with backfill material. However, the method has as many as eight variations, with some of them extracting ore in the underhand direction. The backfilling operation is normally performed after each horizontal slice is removed. An estimated 3% of underground mineral production is derived from cut-and-fill stoping.

The fill material used in this method varies, depending on the support required and the material that may be available to the mine operator. The major types of fill are listed as follows:

(1) Waste fill; (2) Pneumatic fill; (3) Hydraulic fill with dilute slurry; (4) High-density hydraulic fill.

The primary advantage of hydraulic fill is that it can be supplemented with portland cement to allow the fill to harden into a consistency that approaches hard rock. However, dilute hydraulic slurry creates problems because of the large amount of water used. Thus, mining companies often prefer to use high-density fill when possible.

Like other vertical exploration openings, cut-and-fill stopes are generally bounded by pillars for major ground support. Because the stopes are filled, however, these pillars often can be recovered, in part or totally. The timing of fill placement is critical to the success of the method, because the fill must be in place to assume most of the super-incumbent load on the ore in the stope.

Cut-and-fill stoping is a moderate-scale method. The stope must be designed in much the same way as in other vertical stoping methods. Stope dimensions are ordinarily influenced by mechanization factors, such as ease of access, maneuverability of equipment, and production rate requirements. Stope width ranges widely from 6 to 100ft (2 to 30m) and varies with several rock mechanics factors. The smallest equipment available normally sets the minimum width. Stope heights vary from 150 to 300ft (45 to 90m), and stope lengths from 200 to 2000ft (60 to 600m). However, note that the height in the open part of the stope is seldom more than 30ft (10m).

Although many variations of cut-and-fill stoping have been identified, only three will be described in detail here. These are overhand cut-and-fill stoping, drift-and-fill stoping, and underhand cut-and-fill stoping. Each of these is fairly widely practiced, both in the United States and internationally.

An overhand cut-and-fill stoping operation is shown in Fig. 12.1 in the stope area, the miners work under the roof and generally have sufficient head room to move their equipment easily through the stope. The version shown in the figure uses ramps to allow the diesel equipment to move from level to level through the stope area. Maximum grade of the ramps is normally 15% to 20% in the stope area. This version of cut-and-fill stoping is applied in rocks that are relatively strong, which will allow the stope to remain open with only bolting of the roof or bolting combined with mesh. Another variation of this method was widely used in the past, with slushers employed to move the ore. However, diesel equipment has now taken over the method as it offers greater productivity.

Drift-and-fill stoping is a method of mining used for ground conditions that are worse than those for the traditional overhand cut-and-fill stoping. The method can be utilized in either an overhand or an underhand fashion. The mining strategy involves keeping the openings relatively small to reduce the danger of rock failure. Each horizontal slice is removed by drifting forward, using mechanized equipment. After each drift is completed, the drift is backfilled with hydraulic fill to within a few feet (a meter) of the back. This provides excellent support of the hanging wall and footwall rock. A diagram of this method is shown in Fig. 12.2. Note that the vein is four drifts

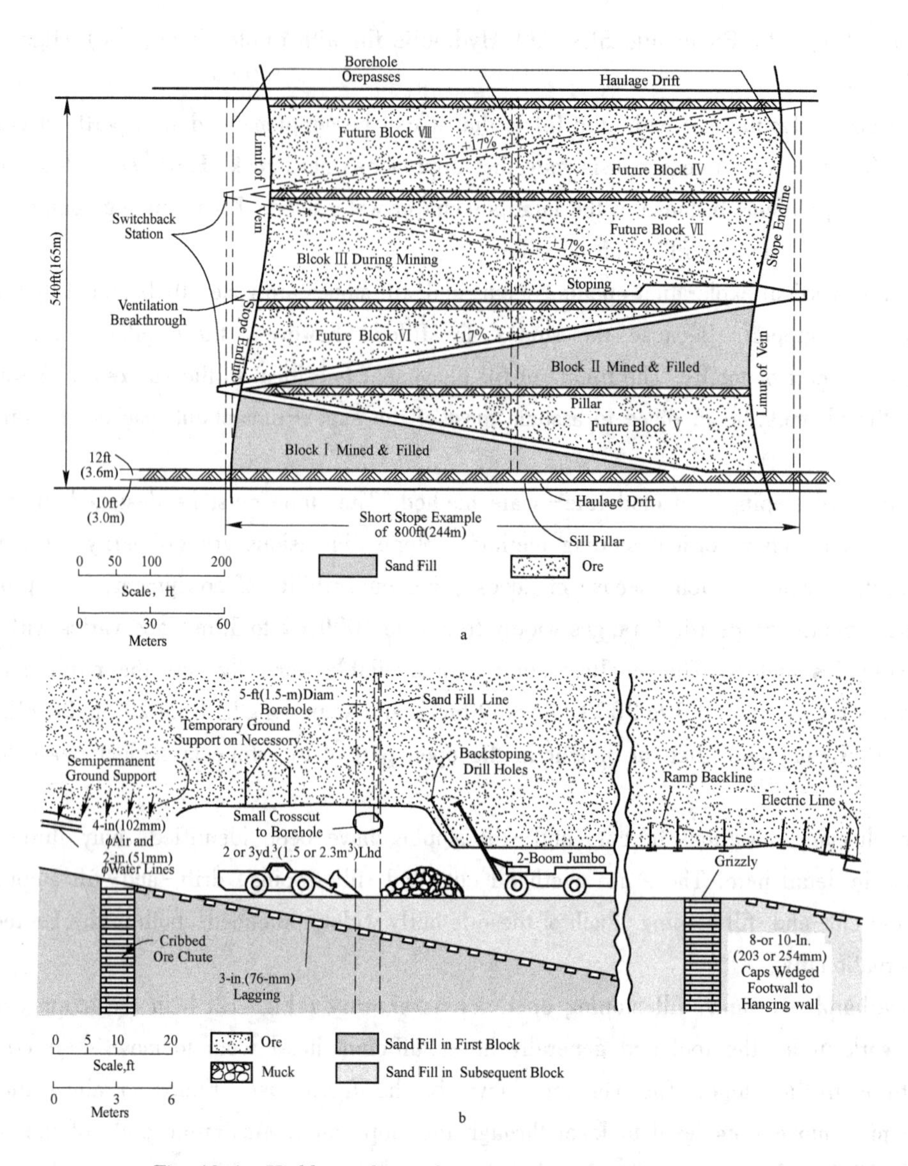

Fig. 12.1 Highly mechanized version of cut-and-fill mining

a—Location of ramps to permit mining of alternate blocks of ore; b—Stoping operations, with drill jumbo and LHD

wide, but that fill is placed after each individual drift is completed. This provides maximum support and minimizes the probability of rock failure.

Underhand cut-and-fill stoping is likewise used in poorer quality rock. It is characterized by the procedure of taking horizontal slices from the top to the bottom of the stope area. To stabilize the surrounding rock after each slice or group of slices, the stope is then filled, usually with cemented hydraulic fill. The fill becomes the roof for subsequent cuts. The method has been practiced with filling placed after every horizontal cut (Fig. 12.3), as well as after a number of cuts have been produced below the fill. The fill is normally a cemented hydraulic mixture to maximize the support capabilities of the material.

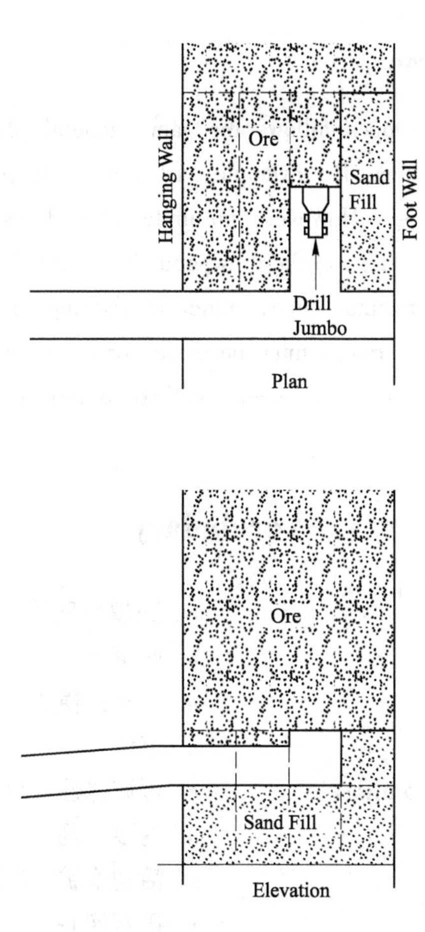

Fig. 12. 2 Drift-and-fill mining using four drifts to span the vein width

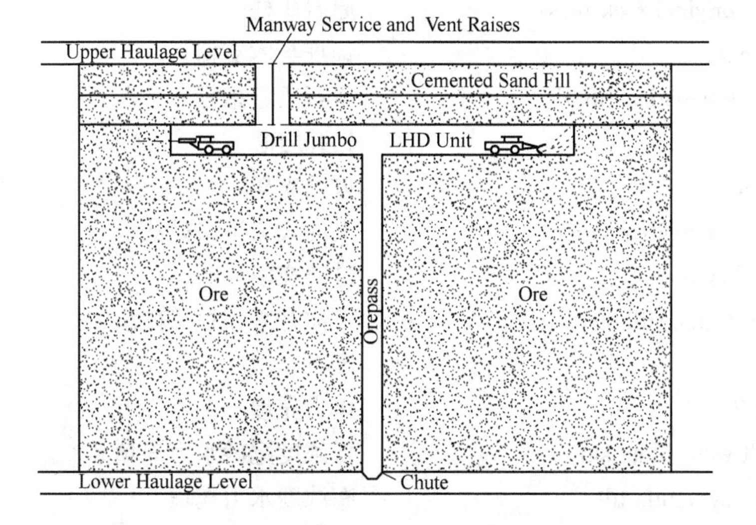

Fig. 12. 3 Longitudinal section of an undercut-and-fill stope

3. Sequence of development

Development of cut-and-fill stopes follows the same general plan as other vertical stoping methods, but there are differences. For example, a haulage drift plus drawpoints and crosscuts are driven for access to the ore. However, an undercut and loading bells are not needed, because the ore moves from the stope to the drawpoints by gravity flow through orepasses. Mining begins by taking the first horizontal slice, either at the top or bottom of stope. If mobile equipment is used in the stope, ramps must be driven or the equipment will be captive in the stope. Some mines use the captive equipment method to reduce the number of development openings that are necessary.

Vocabulary

backfill [ˈbækfɪl]	*vt*. 充填，回填
	n. 回填土
stull [stʌl]	*n*. 支柱；横梁
bound [baʊnd]	*n*. 束缚
maneuverability [məˌnuvərəˈbɪlətɪ]	*n*. 可调遣性；可操纵性；机动性
mesh [meʃ]	*n*. 网眼，网
slusher [sˈlʌʃər]	*n*. 活动拖铲，铲泥机；耙斗
orepass [ɔːrˈpɑːs]	*n*. 矿石溜井
supported methods	充填采矿法
pillars of the original rock mass	原岩矿柱
bolts and timber	锚杆与木支撑
rock quality designation (RQD)	岩石质量指标
drill core	钻孔岩芯
square-setting	方框支架
overhand cut-and-fill stoping	上向式充填采矿法
drift-and-fill stoping	进路充填采矿
underhand cut-and-fill stoping	下向开采充填采矿法
waste fill	废石充填
pneumatic fill	风力充填
hydraulic fill with dilute slurry	低浓度尾砂水力充填
high-density hydraulic fill	高浓度水力充填
portland cement	硅酸盐水泥
grade of the ramp	斜坡道坡度
gravity flow	重力放矿

NOTES

[1] The second category of underground mining methods is referred to as the supported methods. Supported methods are those methods that require some type of backfill to provide substantial amounts of artificial support to maintain stability in the exploitation openings of the mine. Supported methods are used when production openings will not remain standing during their life and when major caving or subsidence cannot be tolerated. In other words, the supported class is employed when the other two categories of methods, unsupported and caving, are not applicable.

第二大类地下采矿法是充填采矿法。充填采矿法是指靠某种形式的充填体作为人工支护，以维持矿山采场稳定。当采场在开采过程中不能保持稳定或地表不允许大范围冒落和沉降，则需要采用充填法进行开采。也就是说，当其他两类采矿方法（空场法和崩落法）不能使用时，则可以采用充填采矿法进行开采。

[2] Pillars of the original rock mass are the ultimate form of ground control in an underground mine because they are capable of providing near-rigid support. In many horizontal methods, natural pillars are relied on for primary support, supplemented by light artificial supports such as bolts and timber. Vertical methods may use natural pillars to surround the stopes, but pillars in the production areas are impractical because they interfere with production operations. Artificial support in the form of pillars of backfill material is therefore used to control the rock mass and make mine output safe and productive.

原生岩体形成的矿柱是地下矿山控制地压的主要形式，因为这些矿柱能提供近似刚性的支护。在许多水平矿床开采过程中，主要靠留下自然矿柱支撑地压，辅助以锚杆或木支护等人工支护手段。在垂直采场中，可以在采场四周留下原生岩体作为矿柱支撑采场地压，但是，在采场内部设置矿柱就不符合实际，因为在采场内部留矿柱会干扰采场生产作业。因此，采用充填材料的人工支护方式就可以控制采场岩体稳定，维持采场生产安全，提高采场生产能力。

[3] In the design of artificial support systems for mining methods, an evaluation, preferably quantitative, of the load-carrying capacity of the natural rock structure is a prerequisite. Rock mechanics tests are performed to evaluate the structural properties of the rock. However to fully understand the capability of the rock to sustain a load, it may be better to determine the rock quality designation (RQD) based on drill core evaluation. The supported class of methods is intended for application to rock ranging in competency from moderate to incompetent.

人工支护系统设计过程中，原岩矿柱的承载能力（最好是定量计算）是设计的基础条件。进行岩石力学测试可以了解岩体结构的承载能力。要很好了解岩体的承载能力，还需要通过钻取岩芯，确定岩体质量指标（即 RQD 值）。充填采矿法适用于稳定性差至中等稳固的矿体。

[4] Cut-and-fill stoping is normally used in an overhand fashion. The ore is extracted in horizontal slices and replaced with backfill material. However, the method has as many as eight variations, with some of them extracting ore in the underhand direction. The backfilling operation is normally performed after each horizontal slice is removed. An estimated 3% of underground mineral production is derived from cut-and-fill stoping.

充填法通常采用上向式开采方式。水平分层开采矿体，之后用充填材料充填采空区。但是，该采矿法有多达八种不同的变化形式，其中有些变化形式就有下向式开采。充填作业往往是在水平分层采完后进行。充填采矿法开采的矿石量占整个地下开采的矿石量的3%。

[5] The fill material used in this method varies, depending on the support required and the material that may be available to the mine operator. The major types of fill are listed as follows: (1) Waste fill; (2) Pneumatic fill; (3) Hydraulic fill with dilute slurry; (4) High-density hydraulic fill.

这种方法中所用的充填材料各种各样，根据所需要的支护等级和矿山周边所具有的充填材料确定充填材料。主要的充填方式有：(1) 废石充填；(2) 风力充填；(3) 低浓度尾砂水力充填；(4) 高浓度水力充填。

[6] The primary advantage of hydraulic fill is that it can be supplemented with portland cement to allow the fill to harden into a consistency that approaches hard rock. However, dilute hydraulic slurry creates problems because of the large amount of water used. Thus, mining companies often prefer to use high-density fill when possible.

水力充填的主要优点是充填体可以添加水泥，使充填体凝结硬化，形成一个整体，接近岩石一般的坚硬。可是，采用低浓度充填时，需要大量水，从而带来一系列问题。因此，矿山公司都尽可能采用高浓度进行充填。

[7] Like other vertical exploration openings, cut-and-fill stopes are generally bounded by pillars for major ground support. Because the stopes are filled, however, these pillars often can be recovered, in part or totally. The timing of fill placement is critical to the success of the method, because the fill must be in place to assume most of the super-incumbent load on the ore in the stope.

与其他急倾斜采场开采一样，充填采场四周往往留有矿柱，以支撑地压。由于采场进行充填，这些矿柱通常可以部分或全部进行回采。充填时间的确定非常重要，因为，必须及时进行充填，才能保证充填体承担采场内矿体上的荷载。

Unit 13 Caving Methods

崩落法

1. Classifaction of methods

We have concentrated our efforts so far on classes of mining methods that require exploitation workings to be held open, essentially intact, for the duration of mining. If the ore and rock are sufficiently competent, unsupported methods are adequate; if ore and rock are incompetent to moderately competent, then supported methods must be used. We now encounter a class of method in which the exploitation openings are designed to collapse; that is, caving of the ore or rock or both is intentional and the very essence of the method.

We define caving methods as those associated with induced, controlled, massive caving of the ore body, the overlying rock, or both, concurrent with and essential to the conduct of mining. There are three current methods that are considered to be caving methods: (1) Longwall mining; (2) Sublevel caving; (3) Block caving.

Longwall mining is used in horizontal, tabular, deposits, mainly coal; the other methods have application to inclined or vertical massive deposits, almost exclusively metallic or nonmetallic.

Because the exploitation openings are deliberately destroyed in the process of mining, the caving class is truly unique. Rock mechanics principles are applied to ensure that caving, in fact, does occur rather than to prevent the occurrence of caving. In effect, the cross-sectional shape of the undercut area (i. e. the width to height, or R/h, ratio) is sufficiently elongated to cause failure of the roof or back. Further, development openings must be designed and located to withstand shifting and caving ground, as well as subsidence that usually extends to the surface. Production must maintained at a steady, continuous pace to avoid disruptions or hangups in the caving action. Good mine engineering and supervision are indispensable to a successful caving operation.

2. Longwall caving

Longwall mining is an exploitation method used in flat-lying, relatively thin, tabular deposits in which a long face is established to extract the mineral. The layout of five longwall panels is illustrated in Fig. 13. 1. Note that the face in Panel 2 is established between the headgate entries and the tailgate entries and that the face is being advanced toward the main entry set (i. e. in the retreat direction). The longwall is kept open by a system of heavy-duty, powered, yielding supports that form a cantilever or umbrella of protection over the face. As a cut or slice is taken along the length of the wall, the supports are collapsed, advanced closer to the face, and reengaged, allowing the roof to cave behind. The caved area is called the gob. In Fig. 13. 1, all of Panel 1 and small portion of Panel 2 have been mined and converted to gob.

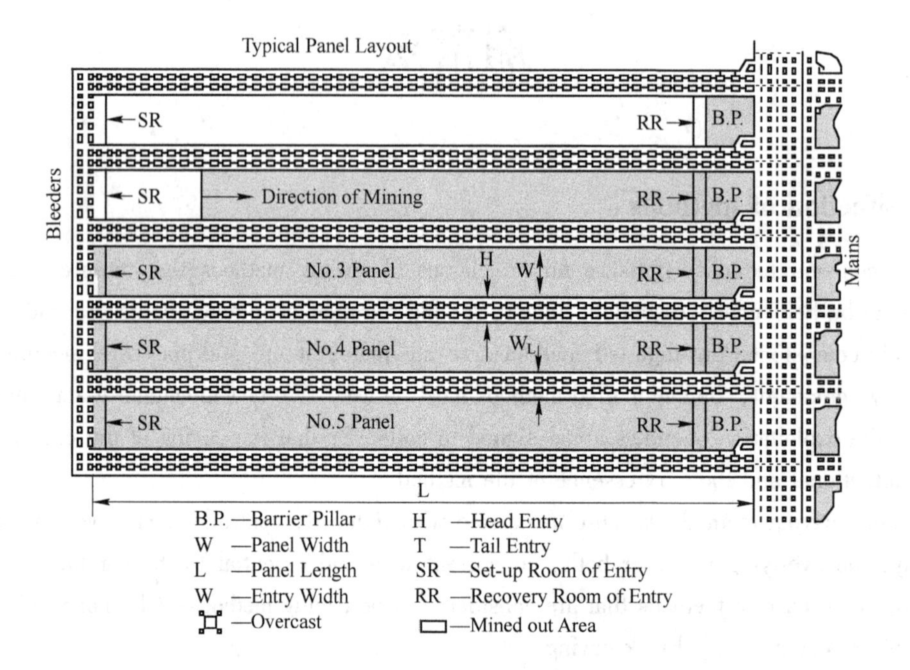

Fig. 13. 1 General layout of a longwall operation

3. Sublevel caving

The two remaining caving methods (sublevel caving and block caving) are applicable to near-vertical deposits of metals or nonmetals. In sublevel caving, overall mining progresses downward while the ore between sublevels is broken overhand; the overlying waste rock (hanging wall or capping) caves into the void created as the ore is drawn off. Mining is conducted on sublevels from development drifts and crosscuts, connected to the main haulage below by ramps, orepasses, and raises. Because only the waste is caved, the ore must be drilled and blasted in the customary way, generally fanhole rounds are utilized.

Fig. 13. 2 illustrates sublevel caving in a steeply dipping ore body.

Because the hanging wall eventually caves to the surface, all main and secondary development is located in the footwall. In vertical cross section, the sublevel drifts and crosscuts are staggered so that those on adjacent sublevels are not directly above one another (Fig. 13. 2). Thus, fanholes driven from one sublevel penetrate vertically to the second sublevel above. Development and exploitation operations are of necessity carefully planned so that adjacent sublevels are engaged in sequential unit operations, as shown in Fig. 13. 2.

Modern sublevel caving bears little resemblance to sublevel caving of yesterday. Formerly a

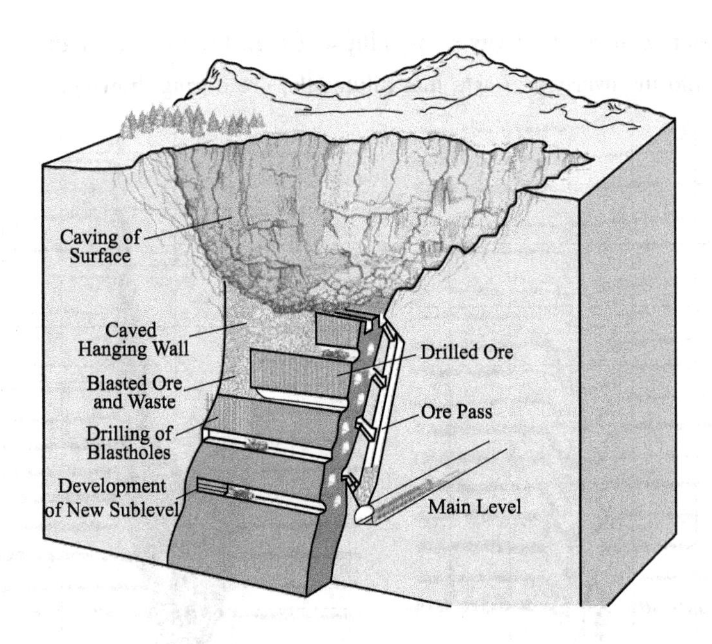

Fig. 13.2 Sublevel caving for a steeply dipping ore body

small-scale, labor-intensive method requiring heavy timbering, sublevel caving evolved through research and development into a highly mechanized, large-scale method of mining using only nominal support. Much of the progress took place in Swedish iron mines; the technology was then exported around the world. In United States, sublevel caving is not widely used. However, in Sweden, Canada, and Australia, the method finds more extensive use because ore deposits in those countries are more amenable to sublevel caving.

Design parameters in sublevel caving are largely a function of caving mechanics, the branch of rock mechanics related to the breakage and collapse of consolidated materials in place and their flow downward by gravity. Although the ore has to be drilled and blasted in sublevel caving, the overlying rock forming the capping or hanging wall is undercut and caves. Extremely careful controls must be exercised in drawing the ore to avoid excessive dilution. Draw control is the practice of regulating the withdrawal of ore in the sublevel crosscuts so as to optimize the economics of the draw. Premature cutoff results in poor recovery, and delayed cutoff produces excessive dilution of the ore. Generally, a cutoff grade based on economics is employed to determine when the mucking should cease and the next fan pattern of holes should be blasted.

Gravity flow of bulk materials has been studied and analyzed in bins, silos and chutes. The results are now being applied to sublevel caving and to specifying its geometry, dimensions and layout. Models are very useful to demonstrate flow principles and have been successful in simulating gravity flow in various caving methods. The simplest are two-dimensional, consisting of two vertical, parallel glass plates filled with horizontal layers of colored sand; as the sand flows by gravity through an opening at the bottom, the layers distend and reveal the flow pattern (Fig. 13.3a). As more sand is withdrawn, the ellipse forms (in three dimensions, it becomes an ellipsoid); we refer to it as a gravity-flow ellipse (Fig. 13.3b). In actual caving, it is this funneling action into the overlying waste that dilutes the ore during drawing.

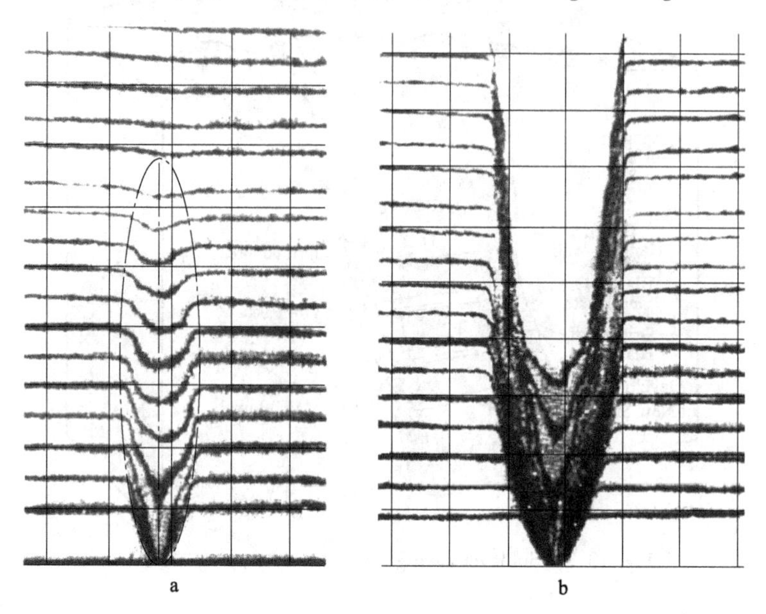

Fig. 13.3 Model studies representing successive phases of bulk-material flow in sublevel caving
a—Vertical ellipse formed at the boundary of gravity motion, delineating the active zone above each crosscut;
b—Advanced phase of gravity flow in drawing ore

The dimensions used in the extraction of ore by sublevel caving have evolved over the years into larger, more efficient values. In early days of mechanized sublevel caving, the sublevel spacings were as close as 30ft (9m) and the crosscut spacings were about 20ft (6m).

However, improved understanding of the flow patterns associated with sublevel caving has resulted in current spacings of 94ft (28.5m) between sublevels and 82ft (25m) between crosscuts. Thus, the changes in the spacings have greatly improved the ratio of production tonnage to development tonnage. Other improvements have also improved the situation in sublevel caving. Better utilization of larger diesel equipment, increased use of remote control in the crosscuts, and automation of some of the haulage operations have all been influential in the quest for productivity improvement.

4. Block caving

The one underground mining method that has the potential to rival surface mining in output and cost is block caving. Block caving is the mining method in which masses, panels, or blocks of ore are undercut to induce caving, permitting the broken ore to be drawn off below. If the deposit is overlain by capping or bounded by a hanging wall, it caves too, breaking into the void created by drawing the ore (Fig. 13.4). This method is unlike sublevel caving, in that both the ore and the rock are normally involved in the caving. As in sublevel caving, the caving proceeds in a columnar fashion to the surface. The result is massive subsidence, accompanied by the exceptionally high production rate and great areal extent characteristic of block caving.

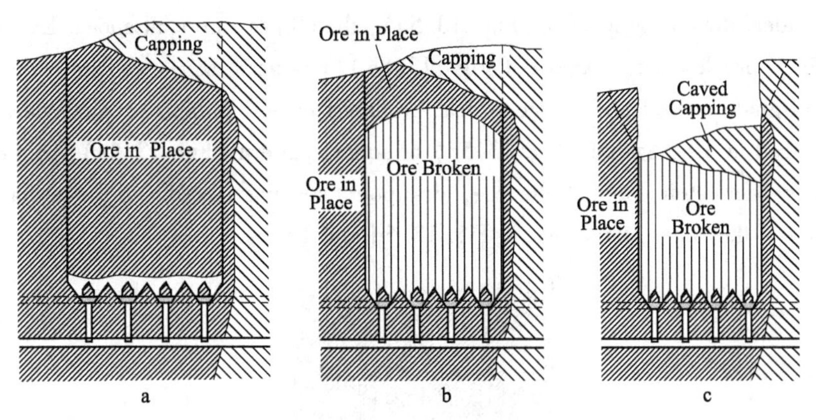

Fig. 13.4 Progressive stages (a, b, c) of block caving, showing caving of ore and waste

A truly American mining method, block caving was invented in the years following World War I to cope with the exploitation of the massive low-grade copper porphyry deposits of the southwestern United States. The technology rapidly spread to the rest of the world and is now applied in many countries. A large-scale method, block caving is utilized to produce about 10% of U. S. underground metals and nonmetals and 3% of all underground minerals.

As shown in Fig. 13.4, the area and volume of ore removed at the bottom of the block during undercutting must be sufficiently large to induce caving in the mass above, which then continues progressively on its own. Steady drawing of the caved ore from the underside of the block provides space for more broken ore to accumulate and causes caving action to continue upward until all the ore in the original block has been caved.

Caving mechanics provides the basis for understanding and controlling the operating factors in block caving, as it does in sublevel caving. Determining the cavability of an ore body is the first task to be undertaken. Good caving action generally requires that the ore body have fractures in three orientations. To investigate the cavability of the ore body, drill cores are obtained throughout the ore body using exploration openings. These cores are then often subjected to rock quality designation (RQD) analysis, which measures the percentage of intact (>4in. or 102mm) core recovered from a drillhole. The RQD value will help to identify the caving characteristics of the rock mass. The RQD values and other methods of determining the suitability of an ore body to caving have been discussed.

Cavability is not just a matter of achieving acceptable fragmentation and optimum operating costs. From a safety standpoint, the ore or capping must not arch over long distances for long periods of time. The formation of stable arches not only disrupts the caving operation but very likely will cause air blast and concussion in the mine when they suddenly collapse.

After cavability is determined for the ore body, the second application of caving mechanics is in draw control and drawpoint spacing. Involved in this problem is the gravity-flow ellipse or ellipsoid referred to in Fig. 13.5, modified by the inflow of waste as the caving funnel progresses upward into the capping. Fig. 13. 5 depicts caving action as a function of drawpoint spacing. With theoretically ideal drawpoint spacing (Fig. 13. 5a), the ellipses are contiguous. Excessive spacing (Fig. 13. 5b) or deficient spacing (Fig. 13. 5c) produces zones of draw that may yield unsatisfactory grade control and create weight problems on sill pillars. In plan view, drawpoints may be arranged in a hexagonal (Fig. 13. 5d) or square pattern (Fig. 13. 5e). To ensure that the zones of draw completely blanket the ore, drawpoint spacing is reduced somewhat, permitting minor overlap of the zones, as shown in Fig. 13. 5d and 13. 5e.

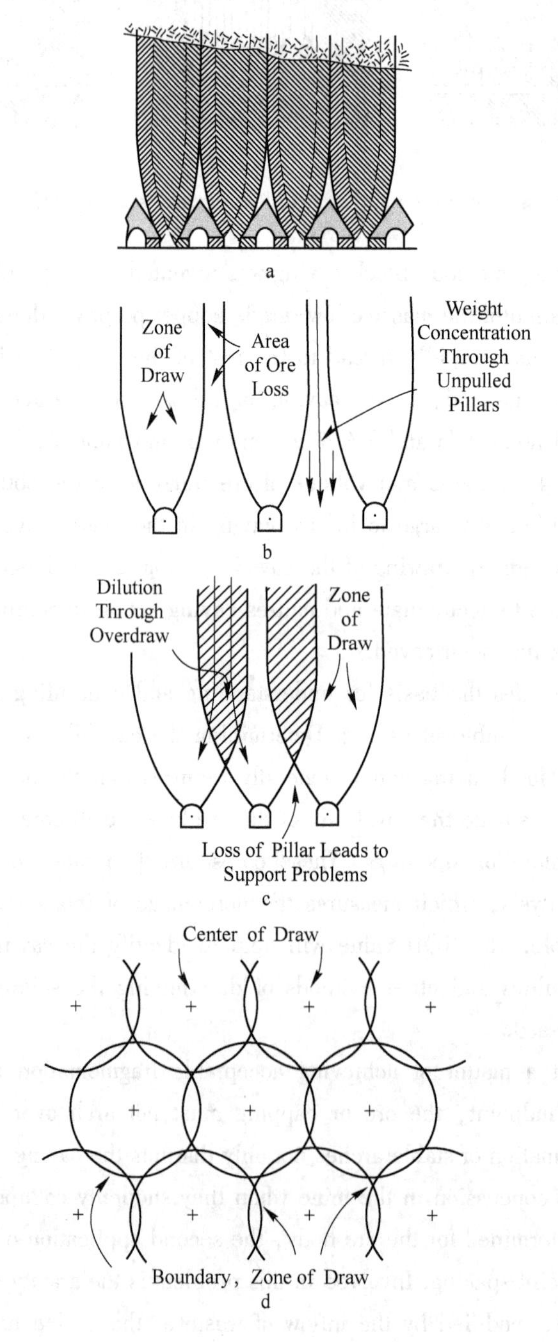

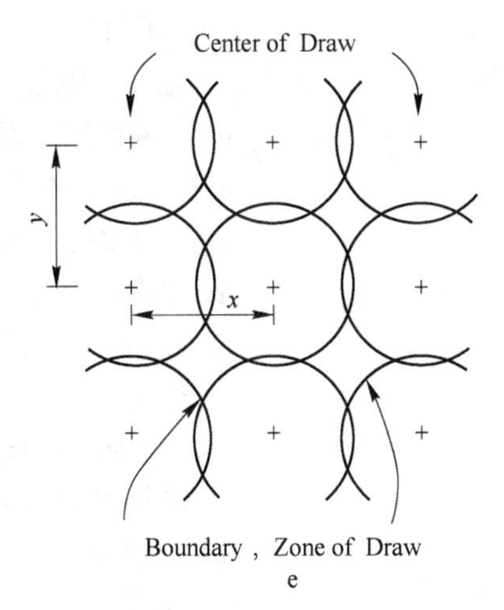

Fig. 13. 5 Effect of drawpoint spacing on zone of draw in block caving. Vertical section
a—Theoretical ideal spacing; b—Excessive spacing; c—Insufficient spacing. Plan views;
d—Suitable hexagonal spacing; e—Suitable square spacing

As the caving actions is initiated, whether in a block, panel, or mass, the caving commences as the undercutting of a critical-sized area is completed and then progresses upward; it also progresses in a planned pattern across the ore body. Once caving begins, the only means of regulating it, as well as the production of ore, is through draw control. Because of irregularities in the contact between ore and waste and, more important, because of funneling in the ore, some dilution is inevitable if a high recovery is achieved. Effective draw control optimizes the combination of grade control, recovery, and dilution. This is illustrated in Fig. 13. 6, where caving action across an ore body is shown. Caving is just being initiated by undercutting on the right while full production is being realized on the left. By practicing proper draw control, the interface between the ore and the rock is kept as intact as possible, minimizing dilution and maximizing recovery and grade. Proper drawpoint spacing will make this possible.

The most critical spacings to be designed into the block caving operation are the drawpoint spacings and the spacings between levels. The spacings between levels dictates how high a column of ore is to be extracted through the drawpoints. In past decades, the drawpoint spacings seldom exceeded 40ft (12m) and the columns were generally less than 400ft (122m) high. However, the tendency is to design the mining system with bigger blocks of ore to be drawn from each drawpoint. The most recent plan at the Henderson Mine is to use 60ft (18m) by 50ft (15m) spacings on the drawpoints and a column height of 800ft (244m). this reduces the development costs and allows a reduction in the overall mining cost if the plan is successful.

Vocabulary

shift [ʃɪft] *n.* 变换；轮班；转移

disruption [dɪsˈrʌpʃən] *n.* 分裂，瓦解

indispensable [ˌɪndɪˈspensəbl] *adj.* 不可缺少的；必需的
cantilever [ˈkæntɪliːvə(r)] *n.* 悬臂；电缆吊线夹板
gob [gɒb] *n.*（采矿的）填塞材料
overhand [ˈəʊvə(r)hænd] *adj.*, *adv.* 举手过肩的（地）
void [vɔɪd] *adj.* 无效的；空虚的
n. 空虚；空处
vt. 使放出；排泄
stagger [ˈstægə(r)] *n.* 交错
sequential [sɪˈkwɛnʃəl] *adj.* 连续的；结果的
premature [ˌpriməˈtjʊr] *n.* 过早发生的事物
adj. 过早的
muck [mʌk] *n.* 岩石
distend [dɪˈstend] *v.*（使）扩张；（使）膨胀；（使）肿胀
ellipse [ɪˈlɪps] *n.* 椭圆（形）
ellipsoid [ɪˈlɪpsɔɪd] *n.* 椭圆体的面
adj. 椭球的
cavability [kəvəˈbɪlɪtɪ] *n.* 可崩性
concussion [kənˈkʌʃən] *n.* 冲击；震动
contiguous [kənˈtɪgjʊəs] *adj.* 接触的；邻近的
hexagonal [heksˈægənl] *adj.* 六边的；六角形的
blanket [ˈblæŋkɪt] *n.* 羊毛毯
adj. 综合的；总括的
columnar [kəˈlʌmnə] *adj.* 圆柱的；柱状的
areal [ˈeərɪəl] *adj.* 地区的；面积的；广大的
caving methods 崩落法
overlying rock 上覆岩层
concurrent with 与……同时发生
longwall mining 长臂式崩落法
sublevel caving 分段崩落法
block caving 阶段崩落法
cross-section 横剖面
roof or back 顶板
avoid disruptions 避免中断，连续
headgate entry 入口
tailgate entries 出口
retreat direction 后退式
fanhole round 扇形炮孔
steeply dipping ore body 急倾斜矿体
premature cutoff 截止出矿过早

delayed cutoff	截止出矿过晚
cutoff grade	截止出矿品位
gravity-flow ellipse	放出椭球体
air blast	空气冲击波
inflow of waste	废石混入
the ratio of production tonnage to development tonnage	采切比

NOTES

[1] We have concentrated our efforts so far on classes of mining methods that require exploitation workings to be held open, essentially intact, for the duration of mining. If the ore and rock are sufficiently competent, unsupported methods are adequate; if ore and rock are incompetent to moderately competent, then supported methods must be used. We now encounter a class of method in which the exploitation openings are designed to collapse; that is, caving of the ore or rock or both is intentional and the very essence of the method.

之前，我们讨论的采矿方法都需要开采空间，在回采期间，这些开采空间必须稳定。如果矿石和岩石稳定性好，就可以采用空场法回采。如果矿岩不稳固至中等稳固，则采用充填法。现在，我们讨论另一类采矿法，其设计的回采空间需要崩落，即人为崩落矿石或岩石或崩落矿岩，而且崩落矿岩是该采矿法的核心。

[2] We define caving methods as those associated with induced, controlled, massive caving of the ore body, the overlying rock, or both, concurrent with and essential to the conduct of mining. There are three current methods that are considered to be caving methods: (1) Longwall mining; (2) Sublevel caving; (3) Block caving.

矿体或上覆岩层岩发生诱导崩落，控制崩落，大规模崩落，在矿岩崩落的同时进行采矿的采矿方法称为崩落法。目前有三种崩落法：（1）长臂式崩落法；（2）分段崩落法；（3）阶段崩落法。

[3] Longwall mining is used in horizontal , tabular, deposits, mainly coal; the other methods have application to inclined or vertical massive deposits, almost exclusively metallic or nonmetallic.

长臂式崩落法常常用于开采水平层状矿体，主要用于开采煤矿。另外两类崩落法用于倾斜或垂直矿体，几乎都是在金属矿床或非金属矿床应用。

[4] Because the exploitation openings are deliberately destroyed in the process of mining, the caving class is truly unique. Rock mechanics principles are applied to ensure that caving, in fact, does occur rather than to prevent the occurrence of caving. In effect, the cross-sectional shape of the undercut area (i. e. the width to height, or R/h, ratio) is sufficiently elongated to cause failure of the roof or back. Further, development openings must be designed and located to

withstand shifting and caving ground, as well as subsidence that usually extends to the surface. Production must maintained at a steady, continuous pace to avoid disruptions or hangups in the caving action. Good mine engineering and supervision are indispensable to a successful caving operation.

若在开采过程中，采场进行人为崩落，则崩落法非常特别。崩落法利用了岩石力学基本原理，即根据岩石力学强度特性，要确保采场矿岩崩落，而不是保护矿岩不崩落。当切割槽的横断面（即切割槽的宽高比）足够大时，采场顶板就会发生崩落。此外，开拓工程及位置的选择必须承受住转移地压或崩落的地压，还要承受住地表沉降。回采中既要保持生产均衡和连续，还要避免生产脱节。若要保持良好的崩落作业，矿山就需要具备优秀的崩落工程师和管理人员。

[5] Longwall mining is an exploitation method used in flat-lying, relatively thin, tabular deposits in which a long face is established to extract the mineral. The layout of five longwall panels is illustrated in Fig. 13. 1. Note that the face in Panel 2 is established between the headgate entries and the tailgate entries and that the face is being advanced toward the main entry set (i. e. in the retreat direction). The longwall is kept open by a system of heavy-duty, powered, yielding supports that form a cantilever or umbrella of protection over the face. As a cut or slice is taken along the length of the wall, the supports are collapsed, advanced closer to the face, and reengaged, allowing the roof to cave behind. The caved area is called the gob. In Fig. 13. 1, all of Panel 1 and small portion of Panel 2 have been mined and converted to gob.

长臂式崩落法用于开采水平薄矿体，在矿体中设立长工作面开采矿石。图 13. 1 描述了具有五个长臂式作业盘区的设计布置。盘区 2 的作业工作面是布置在入口巷道与出口巷道之间，其回采方向是朝主运输巷道推进（即后退式开采）。长臂式作业面靠重型承载梁或保护架提供支撑。当采矿机械沿着工作面进行切割时，去掉远处的支护，在靠近工作面处设置支护，再依次重复作业，切割矿体后，崩落顶板。崩落区域就叫采空区。图 13. 1 中，盘区 1 全区和盘区 2 的小部分区域已经开采完毕成了采空区。

[6] The two remaining caving methods (sublevel caving and block caving) are applicable to near-vertical deposits of metals or nonmetals. In sublevel caving, overall mining progresses downward while the ore between sublevels is broken overhand; the overlying waste rock (hanging wall or capping) caves into the void created as the ore is drawn off. Mining is conducted on sublevels from development drifts and crosscuts, connected to the main haulage below by ramps, orepasses, and raises. Because only the waste is caved, the ore must be drilled and blasted in the customary way; generally fanhole rounds are utilized.

另外两种崩落法（分段崩落法和阶段崩落法）适用于急倾斜金属或非金属矿体的开采。在分段崩落法中，所有分段开采方向是从上而下，矿石在分段的采场内采用上向凿岩爆破落矿，矿体上覆岩层（顶板岩层或围岩）产生崩落，崩落的岩石填补采场放矿时所留下的空间。在采准巷道和切割巷道内进行采矿作业。斜坡道，矿石溜井把采准巷道与主运输道联通在一起。因为只有采场之上的废石产生崩落，因此，采场矿石通常需要进行凿岩、爆破。采场内通常使用扇形孔进行爆破。

[7] Fig. 13. 2 illustrates sublevel caving in a steeply dipping ore body. Because the hanging wall eventually caves to the surface, all main and secondary development is located in the footwall. In vertical cross section, the sublevel drifts and crosscuts are staggered so that those on adjacent sublevels are not directly above one another (Fig. 13. 2). Thus, fanholes driven from one sublevel penetrate vertically to the second sublevel above. Development and exploitation operations are of necessity carefully planned so that adjacent sublevels are engaged in sequential unit operations.

图 13. 2 说明了分段崩落法在急倾斜矿体中的应用。由于矿体上盘岩石最终会崩落到地表，因此，所有主要开拓巷道和次要开拓巷道都位于下盘岩体内。在矿体横剖面中，分段巷道和切割巷道交错布置，因此，相邻两分段中的进路不在一个垂直线上（见图 13. 2）。因此，从一个分段巷道钻凿的扇形炮孔要超过上一个分段的水平标高。分段崩落法的采准作业和回采作业需要认真布置，使得相邻两分段开采衔接顺畅。

Unit 14 Mine Ventilation

矿井通风

1. Introduction

The most vital of the auxiliary operations in underground mining is ventilation. It largely maintains the quality and quantity of the atmospheric environment and is the mainstay of the miner's life-support system and the mine's health and safety program.

The mine ventilation is thc proccss of total air-conditioning responsible for the quantity control of air, its movement, and its distribution. Other processes specifically help to accomplish quality control (e. g. gas and dust control) or temperature-humidity control (e. g. air cooling and dehumidification, heating). Because of its versatility, only ventilation carries out aspects of all three control functions. When the control of the atmospheric environment is complete, that is, when there is simultaneous control of the quality, quantity and temperature, humidity of the air in a designated space, then we are employing total air-conditioning.

In recent years environmental standards in mines have been raised substantially. Although threshold limits are based on human endurance and safety, we are increasingly concerned with standards of human comfort as well. Thereason have to do with cost-effectiveness as well as humanitarianism. Worker productivity, job satisfaction, and accident prevention correlate closely with environmental quality.

Conditioning functions and processes commonly used in mines consist of the following:

(1) Quality control (gas control, dust control);

(2) Quantity control (ventilation, auxiliary or face ventilation, local exhaust);

(3) Temperature-humidity control (cooling and dehumidification, heating).

Processes may be applied individually or jointly. We concentrate here on mine ventilation as the most important and universally used process.

2. Quality control

Chemical contaminants—principally gases and dusts—constitute a variety of hazards in the mine atmosphere. Gases may be suffocating, toxic, radioactive, or explosive. Dusts may be nuisance, pulmonary, toxic, carcinogenic, or explosive. Contaminants occur naturally (e. g. strata gases such as methane in coalbeds or radon gas from radioactive ores) or are introduced by mining activity (e. g. diesel or blasting fumes, smoke from fires, all dusts). Even human breathing liberates a contaminant (carbon dioxide) while consuming oxygen, admittedly in small amounts.

The engineering principles of mine air quality control, for both gases and dusts, are as follows:

(1) Prevention or avoidance.

(2) Removal or elimination.

(3) Suppression or absorption.

(4) Containment or isolation.

(5) Dilution or reduction.

Different practices are employed to implement these control principles. For example, in the room-and-pillar mining of coal with continuous equipment, the two major contaminants are methane gas and coal dust. In this case, corresponding to the five control principles, the following control practices would be considered for application (Table 14. 1):

Table 14. 1 Five control principles for dust and gas dilution

gas	dust	both
1. Nonsparking bits Airway sealent	Sharp bits	Good mining practices Explosion-proof atmosphere
2. Methane drainage	Dust collector Good housekeeping	bleeders
3. Sprayfans	Water sprays Rock dusting	Water infusion
4. Sealing old workings	Hood enclosure	Separate split of air
5. Main ventilation Auxiliary ventilation	Main ventilation Auxiliary ventilation	Main ventilation Auxiliary ventilation

Although not all these principles and practices would be utilized in every mine, enough of them would be employed to cope with exiting hazards and contain them within threshold limit values promulgated by the Mine Safety and Health Administration (MSHA) and other health and safety authorities. Ideally , they should be practices in the order in which they are given; this results in the best quality control and cost-effectiveness. Note that ventilation is listed last, not because it is the least effective but because it is universally applicable and most effective when coupled with other measures.

The quantity of ventilation, Q, required to dilute an airborne hazard is determined by the following relation:

$$Q = \frac{Q_g(1 - TLV)}{TLV - B_g}$$

Where, Q_g is contaminant inflow rate; B_g is concentration of contaminant in normal intake air, and TLV is threshold limit value of the contaminant.

3. Quantity control

Quantity control in mine ventilation is concerned with supplying air of the desired quality and in the desired amount to all working places throughout the mine. Air is necessary not only breathing, a remarkably low 20ft^3/min (0.01m^3/s) per person usually suffices, but to disperse chemical and physical contaminants (gases, dust, heat, and humidity) as well. Because breathing requirements are easily met, federal and/or state laws provide for a higher minimum quantity of air per person (100 to 200ft^3/min, or 0.05 to 0.09m^3/s), a minimum quantity at the face (3000ft^3/min, or 1.4m^3/s) or in the last crosscut of a coal mine (9000ft^3/min, or 4.3m^3/s), or a minimum velocity at the coal face (60ft/min, or 0.3m/s) (Code of Federal Regulations, 2000). Mine ventilation practice is heavily regulated in the United States as well as in the rest of the world, especially in coal and gassy (noncoal) mines, and other statutes relate to air quantities to dilute diesel emissions, blasting fumes, radiation, dusts, and many other contaminants.

Rather surprisingly, legal requirements for minimum airflow seldom govern in determining the design quantity for the mine. Most mines operate with ventilation quantities far in excess of those legislated; the best-ventilated mines circulate millions of ft^3/min (thousands of m^3/s) of air, attaining ratios of the weights of airflow to mineral produced of 10 to 20 tons/ton.

The design basis for specifying the amount of airflow in working places is usually the critical velocity or quantity required to disperse of dilute contaminants (airflow must be well into the turbulent range). If the quantity needed for dilution is inadequate for effective dispersion or cooling, then the velocity and hence the quantity is increased to an adequate level. Economics plays a role also because excessive airflows squander fan horsepower. Critical velocities at the working face range from 100 to 400ft/min (0.5 to 2m/s), unless cooling is a consideration, in which case velocities may attain 400 to 600 ft/min (2 to 3m/s). Knowing the area A and velocity v, the quantity can then be determined: $Q = v \times A$.

Theoretically, contamination by the CO_2 in human exhalation ought to be considered too, but because it is so small ($Q_g = 0.1$ft^3/min, or 47×10^{-6}m^3/s, person), it can be neglected.

Once quantity requirements in all the working places have been specified, it is then necessary to create a pressure difference in the mine to provide the desired flows. Either natural ventilation, in which the pressure difference results from thermal energy (like the chimney effect), or mechanical ventilation, in which the rotational energy of a fan is converted to fluid-flow energy, may serve as an energy source. In modern mines, only fans are relied upon to provide a pressure difference for ventilation (natural ventilation is too variable, unreliable, and insufficient in magnitude to be utilized). Both centrifugal and axial-flow fans are in common use, and they may be installed in either the blower or exhaust position (except that underground booster fans are prohibited in coal mines).

Mine ventilation circuits, like electrical circuits, are arranged with airways in series or parallel or as combination series-parallel circuits called networks. Simple circuits can be solved mathematically, using principles derived from electrical theory; but because of their complexity, networks are best solved by computer and are beyond the scope of this discuss.

4. Temperature-humidity control

In temperature-humidity control, we are concerned with the physical quality of air and its heat content. Both sensible and latent heat are involved because normal air is a mixture of dry air and water vapor. Two air-conditioning processes are employed in mines: (1) heating; (2) cooling and dehumidification. The latter is the more complicated and costly process. Ventilation alone is adequate for cooling until the mine air wet-bulb temperature substantially exceeds 80℉ (32℃), human exertion is seriously impaired and work output and safety suffer. Artificial cooling supplements ventilation in very hot environments. Chilled water is prepared on the surface in cooling towers or underground by refrigeration, and then in heat exchangers (coils or sprays) it cools and dehumidifies the air going to the working faces. Heat is the ultimate constraint in mining; it increases inexorably with depth. Major sources of heat in mines are autocompression (more than 5℉/1000ft of depth, or 10℃/1km) and wall rock or geothermal gradient (0.5~3℉/100ft of depth, or 0.9~5.5℃/100m) groundwater, machinery, human metabolism, blasting, and oxidation may also contribute heat. Although temperature-humidity control applications in mines are still relatively uncommon, they can be expected to increase as underground mines deepen.

5. Summary

The broad objective of mine ventilation and air-conditioning, to provide a comfortable, safe atmospheric environment for workers, should never be neglected. However, its role must be placed in the proper perspective; atmospheric control is only one phase of a broader, more general, environmental-control mission. Mine environmental control can best be administered by a single department that includes health and safety engineering as well, organized in the role of staff to the chief operating officer.

Vocabulary

threshold [ˈθrɛʃhold] n. 开端；入口；门槛；开始

dehumidification [ˈdi:hjʊ(:)mɪdɪfɪˈkeɪʃən] n. 除湿

humanitarianism [hjuˌmænəˈtɛrɪənɪzəm] n. 人类；人性

suffocate [ˈsʌfəˌket] vt. 使窒息
vi. 窒息；呼吸困难

pulmonary [ˈpʌlməˌnɛrɪ] adj. 肺的；有肺病的

carcinogenic [ˌkɑ:sɪnəˈdʒenɪk] adj. 致癌的

bleeder [ˈbli:də(r)] n. 易出血的人；排水管

sprayfan [spˈreɪfæn] n. 喷雾风机

promulgate [prəˈmʌlget] vt. 宣布；传播；颁布；散布

statute [ˈstætʃʊt] n. 成文法；法规；条例

squander [ˈskwɒndə(r)]	vt. 浪费；使分散 n. 漂泊
exhalation [eksəˈleɪʃ(ə)n]	n. 呼出；呼气
blower [ˈbləʊə(r)]	n. 鼓风机
autocompression [ˈɔːtəkəmˈprɛʃən]	n. 自动压缩
metabolism [məˈtæblɪzəm]	n. 新陈代谢；变形
threshold limit	临界极限
prevention or avoidance	预防或回避
removal or elimination	移除或消除
suppression or absorption	抑制或吸收
containment or isolation	封闭或隔离
dilution or reduction	稀释或减少
nonsparking bit	无火花钻头
airway sealcnt	气道封闭剂
methane drainage	瓦斯排放
dust collector	吸尘器
sealing old workings	封闭老空区
water infusion	注水
hood enclosure	机罩外壳
separate split of air	独立风道
concentration of contaminant	污染物浓度
pressure difference	压差
natural ventilation	自然通风
centrifugal and axial-flow fan	离心式和轴流式风扇
underground booster fans	地下增压风机
exhaust position	排气位置，回风井
temperature-humidity	温湿度
water spray	喷水

NOTES

[1] The most vital of the auxiliary operations in underground mining is ventilation. It largely maintains the quality and quantity of the atmospheric environment and is the mainstay of the miner's life-support system and the mine's health and safety program.

地下开采最重要的辅助作业是通风。通风包括井下空气的数量和质量。通风是矿工维持生命的基础，也是矿工健康与安全的根本保障。

[2] The mine ventilation is the process of total air-conditioning responsible for the quantity control of air, its movement, and its distribution. Other processes specifically help to accomplish

quality control (e. g., gas and dust control) or temperature-humidity control (e. g., air cooling and dehumidification, heating). Because of its versatility, only ventilation carries out aspects of all three control functions. When the control of the atmospheric environment is complete, that is, when there is simultaneous control of the quality, quantity and temperature, humidity of the air in a designated space, then we are employing total air-conditioning.

矿井通风是指井下空气质量控制，空气流动方向和风量分配的过程。通风的其他作用是协助完成空气质量控制（有害气体和粉尘控制）或温度-湿度控制（即空气冷却和除湿，增温）。由于通风具有多方面的作用，因此矿井通风具备三个方面的作用，即可以在矿井设计范围内，同时调节井下空气质量，通风量和温度-湿度。如此，我们就可以做到了对井下空气进行调控。

[3] In recent years environmental standards in mines have been raised substantially. Although threshold limits are based on human endurance and safety, we are increasingly concerned with standards of human comfort as well. The reason have to do with cost-effectiveness as well as humanitarianism. Worker productivity, job satisfaction, and accident prevention correlate closely with environmental quality.

近年来，矿山环境标准大幅提高了。尽管人的承受能力和安全是标准制定的基础，但是我们在制作标准时也要考虑人的舒适性。制定的环境标准与成本效率，人性化有关联。当作业环境越好，则工人生产率越高，满意度越高，也可以避免事故发生。

[4] Conditioning functions and processes commonly used in mines consist of the following:

(1) Quality control (gas control, dust control);

(2) Quantity control (ventilation, auxiliary or face ventilation, local exhaust);

(3) Temperature-humidity control (cooling and dehumidification, heating).

矿山常使用的环境控制措施如下：

(1) 空气质量控制（有害气体控制，粉尘控制）；

(2) 通风量的控制标准（通风量，辅助巷道或作业面的通风量和尾气量排放量）；

(3) 温度湿度控制（冷却和除湿，加热）。

[5] Processes may be applied individually or jointly. We concentrate here on mine ventilation as the most important and universally used process.

这些环境控制措施可以单独应用，也可以联合应用。我们主要强调矿山通风作为环境控制的主要标准。

[6] Chemical contaminant, principally gases and dusts, constitute a variety of hazards in the mine atmosphere. Gases may be suffocating, toxic, radioactive, or explosive. Dusts may be nuisance, pulmonary, toxic, carcinogenic, or explosive. Contaminants occur naturally (e. g. strata gases such as methane in coalbeds or radon gas from radioactive ores) or are introduced by mining activity (e. g. diesel or blasting fumes, smoke from fires, all dusts). Even human breathing liberates

a contaminant (carbon dioxide) while consuming oxygen, admittedly in small amounts.

井下化学污染物主要是指有害气体和粉尘，它们是矿山环境的主要有害物质。有害气体可以令人窒息，中毒，具有放射性，也可能爆炸。粉尘可以令人员作呕，产生尘肺病，中毒，致癌，或产生粉尘爆炸。井下环境污染物可能来源于自然界（即，岩层内有害气体，如瓦斯，或放射性岩石里含有的氡气）或来源于采矿活动（如，柴油机或不爆炸物，火烧烟雾，以及粉尘）。人类呼吸时在某种程度上，也要消耗氧气，产生有害物（二氧化碳）。

[7] The engineering principles of mine air quality control, for both gases and dusts, are as follows:

(1) Prevention or avoidance.

(2) Removal or elimination.

(3) Suppression or absorption.

(4) Containment or isolation.

(5) Dilution or reduction.

控制矿井的有害气体措施（包括有害气体数量和粉尘含量）如下：

(1) 预防和避免有害气体和粉尘的产生。

(2) 消除有害气体和粉尘。

(3) 降低或吸收有害气体和粉尘。

(4) 抑制或隔离有害气体和粉尘。

(5) 稀释或降低有害气体和粉尘。

Unit 15 The History of Surface Mining

露天开采发展历史

1. Introduction

A fascinating thread that runs through the history of mining is the continuing evolution of mining methods. Often, the initial exploitation of a deposit involved rudimentary scratching at outcrops and picking up pieces of ore from the surface. This surface method was then followed in many instances by the development of underground workings in the form of shafts and galleries. Finally, a surface operation, often on a large-scale, would take place.

Mining was the second of man's endeavors, agriculture was the first. Since prehistoric time, mining has been integral and essential to man's existence. The earliest relatively large-scale mining for outcropping native copper occurred between 5000 and 15000 BC. Rock fragmentation was usually achieved by the cyclical application of fire and water. Loading and haulage was performed by manual labor with stone, wooden, and bronze tools for excavation and animals and human beings were used for haulage.

2. Early endeavors

Mining began with Paleolithic man about 450000 years ago. The first use of metals was for decoration rather than for utility purposes because of their unusual character and rarity. The cultural stages of the evolution of man are associated with minerals and are the Stone Age (prior to 4000BC), Bronze Age (4000 to 1500BC), Iron Age (1500BC to 1780AD), Steel Age (1780~1945), and the Nuclear Age (since 1945).

3. Bronze age

Evidence of early copper mining exists in many parts of the world. The advent of both the Bronze and Iron Ages was contingent upon man's discovery of smelting and learning to reduce ores to native metal or alloy form. The art of rock breakage by fire setting was the first technological breakthrough in mining.

4. Iron age

The introduction of iron for making tools and weapons changed the life of early man in a vast number of ways. The earliest objects that have survived were made of meteoric iron, which contains a high percentage of nickel, and were picked up from the ground. The use of iron was made feasible through the development of three processes: "steeling", the addition of carbon to ore, "quenching" the sudden cooling of hot metal, and "tempering" the reheating of quenched metal to correct for brittleness.

5. Technological development

5.1 In precious metals mining

Innovative technology has been the utilization of hydraulic shovels, continuous mining systems, computer-assisted production control and scheduling, and improved bulk materials conveying and handling.

5.2 In industrial minerals mining

Development has been at a brisk level. There have been efforts to incorporate more mine planning and scheduling techniques to improve productivity. Long-term development plans have been devised.

5.3 In leaching

A technological development introduced in the late 1960s, heap leaching, has made significant contributions to the viability of precious metals operations, both for low-grade deposits and the reworking of old properties. The first commercial application of the technology occurred in the late 1960s at Carlin Gold Mining in Nevada.

It is in the past ten years that heap leaching has developed into an efficient method of treating oxidized gold and silver ore. It has proven to be both an efficient way to extract precious metals from small, shallow deposits, as well as an attractive way to treat large, low-grade, disseminated deposits.

The technology is also being used to recover the metal values from waste dumps at old mining properties.

6. Equipment trends

A most important development in surface mining in recent years has been the use of increasingly sophisticated onboard electronics and microcomputer systems for mining equipment. These systems range from those used to assist the personnel who is operating the hydraulic shovel or walking dragline to managing and monitoring the performance and productivity of the mine's mobile equipment.

Large shiftable mining equipment will be necessary in the future as mining of deeper near-surface deposit becomes necessary. This will be particularly necessary where climatic conditions are extreme or overburden removal is difficult.

The development and application of high-angle, elevating, and cross-pit conveyors result in considerable reductions in transportation costs. Conveyors are an alternative to trucks and other diesel-consuming transport. Trucks and conveyors can be combined through using an in-pit crusher. These can be fully mobile, semi-mobile, or fixed crushing plants. One advantage is to reduce material to the size limit transportable by conveyors.

Draglines that are crawler-mounted have resulted in reduced time and cost of erection. A major development has been the availability of long-boom draglines.

Another development has been drill rigs with computer-based programmable controllers.

Other significant developments include renewed interest in trolley-assist, truck dispatch systems, higher horsepower, and greater capacity.

Continuous surface miners have been developed that incorporate a rotating cutter drum and conveyor discharge system. Such surface miners are especially applicable to multiple seam mining where the seams are separated by thin bands of overburden or in cases where seams are split and where materials of different qualities must be separated. Surface miners can mine to very narrow limits, improving resource recovery and providing an uncontaminated product.

Hydraulic excavators are competitive with small to medium-size cable shovels and wheeled front-end loaders, especially in smaller open pits such as those in many of the new gold operations.

Scrapers, front-end loaders and electric cable shovels comprise the traditional equipment for surface mining. All classes of conventional loading equipment have undergone design changes, including improved electrics and incorporation of health monitoring and diagnostic systems. These changes are resulting in increased reliability, improved performance, and a lower unit cost of production.

Dredging is an attractive means of lowering mining costs, particularly in gold operations.

On-board radios and computer equipment are focal points for improved communication and control and mine haulage efficiency.

Other recent developments in surface mining technology include the wider use of emulsion-type explosives and developments already mentioned: increased use of hydraulic excavators, use of in-pit crushers, and shiftable belt conveyors.

7. Future trends

Responding to increased competition, the future of surface mining has become a showpiece for technological innovation to meet sharply rising production costs.

Future trends will see even greater mine productivities as a result of innovations in:

(1) Off-highway truck design and performance.

(2) Hydraulic shovel reliability and durability.

(3) Computer-aided controls in mine operations and design.

(4) Blasting agent utilization and detonation efficiencies.

(5) Management planning and manpower scheduling and utilization.

(6) Continuous mining and materials handling systems.

In closing, surface mining will continue in the forefront of innovation, mandated by the industry's commitment for excellence in cost competition and requirements for meeting world material demands.

Vocabulary

thread [θred]	*n.* 线
	vt. 以线穿过
	vi. 谨慎地穿过
scratch [skrætʃ]	*vt.* 抓；擦伤
	vi. 被抓破
	n. 抓痕
	adj. 偶然的
gallery [ˈgæləri]	*n.* 画廊；走廊；艺术陈列馆；巷道
paleolithic [ˌpæliəuˈliθik]	*adj.* 旧石器时代的
contingent [kənˈtindʒənt]	*adj.* 意外的；偶发的
	n. 可能的事件
smelt [smelt]	*n.* 胡瓜鱼
	v. smell 的过去式及过去分词
	vt. 熔炼
	vi. 被熔炼
reduce [riˈdjuːs]	*vt.* 缩减；减少；使简化；[化学] 使还原
breakthrough [ˈbreikˈθruː]	*n.* 突围；突破
nickel [ˈnikl]	*n.* 镍
	vt. 覆以镍
quench [kwentʃ]	*vt.* 熄灭，扑灭
	vi. 冷却
tempering [ˈtempəriŋ]	*n.* 回火
grinding [ˈgraindiŋ]	*adj.* 磨的
separation [ˌsepəˈreiʃən]	*n.* 分离，缺口，选矿
palletize [ˈpælitaiz]	*vt.* 把……放在托盘上；使成堆
viability [ˌvaɪəˈbɪlətɪ]	*n.* 生存能力
disseminate [dɪˈsemɪneɪt]	*vt.* 传播；散布
dragline [ˈdræglaɪn]	*n.* 牵引绳索 [矿物] 绳头电铲，索头铲
meteoric iron	陨石铁
be contingent upon	以……为条件
hydraulic shovel	液压铲
heap leaching	堆浸
disseminated deposit	浸染状矿床
waste dump	废石场
shiftable mining equipment	移动式采矿设备
near-surface deposit	浅表矿床

truck dispatch system 卡车调度系统
crawler-mounted 安装有行走机构
long-boom dragline 长臂式索斗铲
trolley-assist 助力车
rotating cutter drum 旋转刀盘
conveyor discharge system 传送机排料系统
multiple seam mining 多层矿脉开采

NOTES

[1] A fascinating thread that runs through the history of mining is the continuing evolution of mining methods. Often, the initial exploitation of a deposit involved rudimentary scratching at outcrops and picking up pieces of ore from the surface. This surface method was then followed in many instances by the development of underground workings in the form of shafts and galleries. Finally, a surface operation, often on a large-scale, would take place.

采矿发展的历史脉络也就是采矿方法不断发展演变的过程。通常，最初的矿床开采主要是在地表露头矿攫取，在地表捡拾单个矿块。在多数情况下，地表开采之后，就要发展到地下开采，地下开采需要掘进竖井和平巷。露天开采往往是一种大规模的开采，但是随着开采深度的增加，最终会被地下开采所取代。

[2] Mining was the second of man's endeavors — agriculture was the first. Since prehistoric time, mining has been integral and essential to man's existence. The earliest relatively large-scale mining for outcropping native copper occurred between 5000 and 15000BC. Rock fragmentation was usually achieved by the cyclical application of fire and water. Loading and haulage was performed by manual labor with stone, wooden, and bronze tools for excavation and animals and human beings were used for haulage.

人类最早的实践活动是农业耕作，采矿是人类第二个阶段的实践活动。在史前时代，采矿就是人类活动的基本组成部分。最早相对较大规模的采矿活动是公元前 5000~15000 年的地表露头开采自然铜。通过反复对岩石进行火烧水冷，进而破碎岩石。采用石头、木头和铜制工具进行人力开采矿石，利用动物进行矿石运输。

[3] Mining began with Paleolithic man about 450000 years ago. The first use of metals was for decoration rather than for utility purposes because of their unusual character and rarity. The cultural stages of the evolution of man are associated with minerals and are the Stone Age (prior to 4000BC), Bronze Age (4000 to 1500BC), Iron Age (1500BC to 1780AD), Steel Age (1780~1945), and the Nuclear Age (since 1945).

约 45 万年前的石器时代，就开始有采矿活动。在早期，由于金属具有独特性能，且稀缺，因此，金属最早是作为装饰品，而不是用来制作工具使用。人类文明发展的过程与矿物有关，包括石器时代（公元前 4000 年），铜器时代（公元前 4000 年至公元前 1500

年)，铁器时代（公元前1500年至公元1780年)，钢器时代（公元1780年至1945年)，原子时代（1945年至今)。

[4] Evidence of early copper mining exists in many parts of the world. The advent of both the Bronze and Iron Ages was contingent upon man's discovery of smelting and learning to reduce ores to native metal or alloy form. The art of rock breakage by fire setting was the first technological breakthrough in mining.

很早以前，世界许多地区就存在采铜的历史遗迹。随着人类发明的冶炼，把矿石还原成自然金属或合金，相应地就出现了铜器时代和铁器时代。采矿工程的第一个技术突破是采用火来破碎岩石的技术。

[5] The introduction of iron for making tools and weapons changed the life of early man in a vast number of ways. The earliest objects that have survived were made of meteoric iron, which contains a high percentage of nickel, and were picked up from the ground. The use of iron was made feasible through the development of three processes: "steeling", the addition of carbon to ore, "quenching" the sudden cooling of hot metal, and "tempering" the reheating of quenched metal to correct for brittleness.

随着人类利用铁来制作工具和武器，这样就改变了早期人类的许多生活方式。现存的早期铁制物品是由陨石铁制作而成的，这种铁器中含有高比例的镍，且无须开采，直接从地表捡拾就可以用来加工制作。经过三个加工过程就可以制作铁器物品，即炼钢（在矿石中加入一定的碳)，淬火（热金属突然冷却)，回火（淬火后的金属重新加热，以便减低其脆性)。

Unit 16 Planning and Design of Surface Mines

露天开采计划与设计

1. Introduction

Many factors govern the size and shape of an open pit. These must be properly understood and used in the planning of any open pit operation. The following are the key items affecting the pit design: geology, grade, localization of the mineralization, extent of the deposit, topography, property boundaries, production rates, bench height, pit slopes, road grades, mining costs, processing costs, metal recovery, marketing considerations, strip ratios, cutoff grade.

2. Bench height

The bench height is the vertical distance between each horizontal level of the pit.

The elements of a bench are illustrated in Fig. 16. 1. Unless geological conditions dictate otherwise, all benches should have the same height.

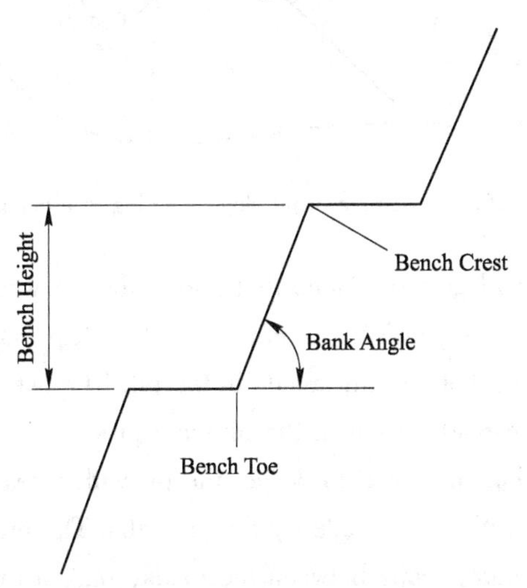

Fig. 16. 1 Bench cross section

The height will depend on the physical characteristics of the deposit, the degree of selectivity required in separating the ore and the size and type of equipment to meet the production requirements and climatic conditions.

The bench height should be set as high as possible within the limits of the size and type of equipment selected for the desired production. The bench should not be so high that it will present safety problems of towering banks of blasted or un-blasted material.

The bench height in open pit mines will normally range from 15m in large copper mines to as little as 1m in uranium mines.

3. Pit slopes

The slope of the pit wall is one of the major elements affecting the size and shape of the pit. The pit slope helps determine the amount of waste that must be moved to mine ore. The pit wall needs to remain stable as long as mining activity is in that area. The stability of the pit walls should be analyzed as carefully as possible. Rock strength, faults, joints, presence of water, and other geologic information are key factors in the evaluation of the proper slope angle.

The physical characteristics of the deposit cause the pit slope to change with rock type, sector location, elevation, or orientation within the pit. Fig. 16. 2 illustrates how the pit slopes may vary in the deposit.

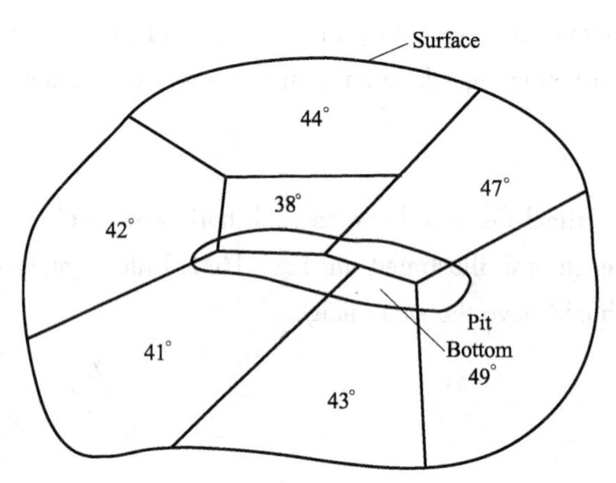

Fig. 16. 2 Example of pit slopes varying in a deposit

A proper slope evaluation will give the slope that allows the pit walls to remain stable. The pit walls should be set as steep as possible to minimize the strip ratio. The pit slope analysis determines the angle to be used between the roads in the pit. The overall pit slope used for design must be flatter to allow for the road system in the ultimate pit.

Fig. 16. 3 and Fig. 16. 4 show the need to design the pit with a lesser slope to allow for roads. Fig. 16. 3 has been designed with a 45° angle for the pit walls. The pit in Fig. 16. 1 uses the same pit bottom and the 45° inter-ramp slope between the roads, but, a road has been added.

Note the larger pit results. In the example, almost 50% more tonnage must be moved to mine the same pit bottom.

In the early design of a pit a lesser pit slope can be used to allow for the road system. The overall slope to use will depend on the width, grade, and anticipated placement of the road.

Fig. 16. 3 Pit designed with a 45° pit slope

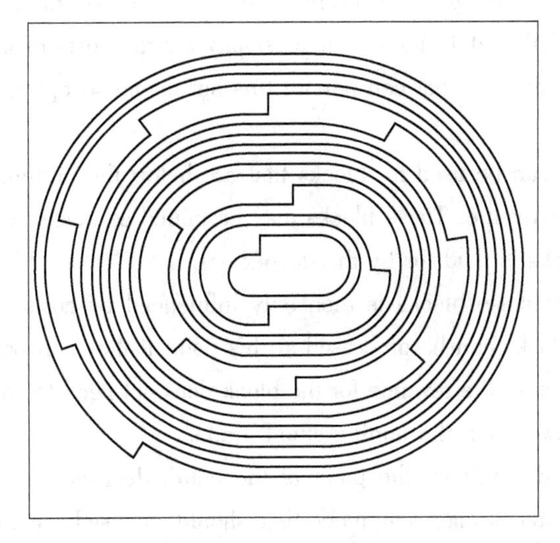

Fig. 16. 4 Pit designed with a 45° interramp slope and a road system

Fig. 16. 5 shows a vertical section of a pit wall from Fig. 16. 4. The inter-ramp angle is projected from the pit bottom upward to the original ground surface at point B.

The overall pit slope angle is the angle from the toe of the bottom bench to the crest of the top bench. Point A shows the intercept of the overall pit slope angle with the original ground surface.

4. Cutoff grade

Cutoff grade is any grade that for any specific reason is used to separate any two courses of action. The reason used in setting a cutoff grade usually incorporates the economic characteristics of the project.

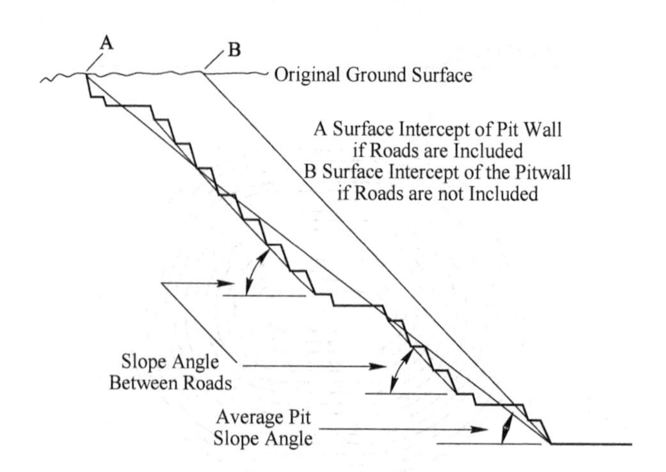

Fig. 16.5 Vertical section through a pit wall

When mining, the operator must make a decision as to whether the next block of material should be mined and processed, mined and stockpiled, mined (to expose ore) and sent to the waste dump, or not mined at all. The grade of the block is used to make this decision.

For any block to be deliberately mined, it must pay for the costs of mining, processing, and marketing. The grade of material that can pay for this but for no stripping is the breakeven mining cutoff grade.

A second cutoff grade can be used for blocks that are below the mining cutoff grade and would not be mined for their own value. These blocks may be mined as waste by deeper ore blocks. The cost of mining these blocks is paid for by the deeper ore.

The final destination of these blocks is then only influenced by costs for the blocks once they have been mined. The blocks can be processed at this point if they can pay for just the processing and marketing costs. Because the revenue for the block does not need to cover the mining cost, the milling cutoff grade is lower than the mining cutoff grade.

The cutoff calculation depends on the point of the cutoff decision in the life of the mine. All direct costs of mining, processing, and marketing should be used. In mining phase this would include the drilling, blasting, loading, and hauling costs.

The processing costs would cover crushing, conveying, grinding, and concentrating costs.

The marketing costs could include concentrate handling, smelting, refining and transportation. Additional direct costs for royalties and taxes would also be included. Overhead costs should also be added to the calculation. Depreciation is used in the calculation for the purpose of setting the pit size.

5. Strip ratio

The strip ratio is the ratio of the number of tons of waste that must moved for one ton of ore to be mined. The results of a pit design will determine the tons of waste and ore that the pit contains. The ratio of waste and ore for the design will give the average strip ratio for that pit. This differs from the breakeven strip ratio used to design the pit.

The breakeven strip ratio refers only to the last increment mined along the pit wall. The strip ratio is calculated for the point at which breakeven occurs and the necessary stripping is paid for by the net value of the ore removed.

The calculation for the breakeven strip ratio (BESR) is

$$BESR=(A-B)/C$$

Where A=revenue per ton of ore;

B=production cost per ton of ore (including all costs to the point of sale, excluding stripping);

C=stripping cost per ton of waste.

In certain studies a minimum profit requirement is included in the formula.

$$BESR=[A-(B+D)]/C$$

Where D=minimum profit per ton of ore.

Vocabulary

grade [greid]	n. 年级；阶段；等级
topography [tə'pɒgrəfi]	n. 地形学，地志，地势
intercept [ˌɪntə'sept]	vt. 中途拦截
	n. 截距；拦截（物）
breakeven [ˌbreɪk'iːvən]	n. 收支平衡
depreciation [dipriːʃi'eiʃən]	n. 通货膨胀
bench height	台阶高度
pit slope	露天坑边坡角
road grade	运输道坡度
strip ratio	剥采比
cutoff grade	边界品位
selectivity required in separating the ore	选别开采
towering banks of blasted material	爆堆
pit wall	露天边帮
ultimate pit	最终境界
inter-ramp slope	出入沟之间的边坡
overall slope	最终边坡
width, grade, and anticipated placement of the road	道路的宽度、坡度、预期的位置
vertical section of a pit wall	露天边坡横剖面
the toe of the bottom bench	坑底台阶坡底线
the crest of the top bench	最高台阶坡顶线
mined and processed	采掘并选矿
mined and stockpiled	采掘并堆存

mined (to expose ore) and sent to the waste dump	采掘（为揭露矿体）并排岩
overhead cost	企业一般管理成本
breakeven strip ratio	经济合理剥采比
long or short-range planning	长期或短期计划
the limits of the open pit	最终境界

NOTES

[1] Many factors govern the size and shape of an open pit. These must be properly understood and used in the planning of any open pit operation. The following are the key items affecting the pit design: geology, grade, localization of the mineralization, extent of the deposit, topography, property boundaries, production rates, bench height, pit slopes, road grades, mining costs, processing costs, metal recovery, marketing considerations, strip ratios, cutoff grade.

露天坑的大小和形状是由许多因素决定的。我们要正确理解这些因素，并利用这些因素来计划露天坑的开采作业。下面这些因素是影响露天坑设计的关键因素：矿床地质、矿石品位，矿床地理位置，矿床范围，地表地形，开采边界，生产能力，台阶高度，露天矿边坡角，道路坡度，采矿成本，加工成本，金属回收率，市场因素，剥采比，边界品位。

[2] The bench height is the vertical distance between each horizontal level of the pit. The elements of a bench are illustrated in Fig. 16. 1. Unless geological conditions dictate otherwise, all benches should have the same height.

台阶高度是露天坑相邻两水平层之间的垂直距离。台阶的要素如图 16. 1 所示。在矿床地质条件没有做出特殊说明时，所有台阶高度都一样。

[3] The height will depend on the physical characteristics of the deposit, the degree of selectivity required in separating the ore and the size and type of equipment to meet the production requirements and climatic conditions.

台阶高度值与矿床的物理特性，选别开采要求，满足产量的设备规模和类型，气候条件有关。

[4] The bench height should be set as high as possible within the limits of the size and type of equipment selected for the desired production. The bench should not be so high that it will present safety problems of towering banks of blasted or un-blasted material.

The bench height in open pit mines will normally range from 15m in large copper mines to as little as 1m in uranium mines.

根据产量要求所选择的设备一定时，露天坑的台阶高度应该尽量高。台阶高度也不能太高，台阶高度太高，爆堆高度就大，从而危及设备安全。

大型露天铜矿的台阶高度可达 15m，而铀矿的台阶高度低至 1m。

[5] The slope of the pit wall is one of the major elements affecting the size and shape of the pit. The pit slope helps determine the amount of waste that must be moved to mine ore. The pit wall needs to remain stable as long as mining activity is in that area. The stability of the pit walls should be analyzed as carefully as possible. Rock strength, faults, joints, presence of water, and other geologic information are key factors in the evaluation of the proper slope angle.

露天矿边坡角是确定露天矿大小和形状的主要因素。露天矿边坡角决定了露天坑内需要剥离的废石量和采出的矿石量。露天矿边坡角在开采期间要保持稳定。露天矿边坡角需要尽量仔细分析研究。边坡岩石的强度、边坡断层、节理、边坡含水情况，以及其他地质因素是确定合理边坡角的主要因素。

[6] The physical characteristics of the deposit cause the pit slope to change with rock type, sector location, elevation, or orientation within the pit. Fig. 16.2 illustrates how the pit slopes may vary in the deposit.

矿床物理特性也引起边坡角发生改变。露天坑内边坡中，不同岩石类型，不同区域，不同标高和不同方向所对应的边坡角也不一样。图 16.2 说明了露天坑边坡的变化情况。

[7] A proper slope evaluation will give the slope that allows the pit walls to remain stable. The pit walls should be set as steep as possible to minimize the strip ratio. The pit slope analysis determines the angle to be used between the roads in the pit. The overall pit slope used for design must be flatter to allow for the road system in the ultimate pit.

合适的边坡角可以保证边坡稳定。边坡角应该尽可能陡，使剥采比达到最小。边坡稳定性分析决定了开拓坑线之间的那段边坡的角度值。露天坑设计的总边坡角必须比这个开拓坑线之间的边坡角要小，以便在最终境界上设计运输坑线。

[8] Fig. 16.3 and Fig. 16.4 show the need to design the pit with a lesser slope to allow for roads. Fig. 16.3 has been designed with a 45° angle for the pit walls. The pit in Fig. 16.4 uses the same pit bottom and the 45° inter-ramp slope between the roads, but, a road has been added.

图 16.3 和图 16.4 表明为了在边坡上布置运输坑线，露天实际边坡角要比稳定边坡角小。图 16.3 中设计的边坡角为 45°。图 16.4 为同一个露天坑底，运输坑线之间的边坡角为 45°，但是，在边坡中增加的运输坑线。

[9] Note the larger pit results. In the example, almost 50% more tonnage must be moved to mine the same pit bottom.

因此，增加了运输坑线之后，露天坑增大了。在这个实例中，开采相同矿石，其剥离的废石增加了约 50%。

[10] In the early design of a pit a lesser pit slope can be used to allow for the road system. The

overall slope to use will depend on the width, grade, and anticipated placement of the road. Fig. 16. 5 shows a vertical section of a pit wall from Fig. 16. 4. The inter-ramp angle is projected from the pit bottom upward to the original ground surface at point *B*.

在早期的露天坑设计中，为了在边坡中布置运输坑线，其边坡角要减小。总边坡角大小与运输坑线宽度，坡度，坑线设计位置有关。图 16. 5 为图 16. 4 的横剖面。沿坑线之间的边坡角，从坑底投射到原始地面后是 *B* 点。

[11] The overall pit slope angle is the angle from the toe of the bottom bench to the crest of the top bench. Point *A* shows the intercept of the overall pit slope angle with the original ground surface.

总边坡角是最下一个台阶的坡底线至最上一个台阶的坡顶线之间的夹角。最下一个台阶的坡底线至地面 *A* 点的边坡中包含了运输坑线。

[12] Cutoff grade is any grade that for any specific reason is used to separate any two courses of action. The reason used in setting a cutoff grade usually incorporates the economic characteristics of the project.

边界品位是指区分采矿作业还是剥离废石作业的品位，大于这个品位的开采是采矿作业，小于这个品位的开采是剥离作业。确定边界品位必须考虑经济性原则。

[13] When mining, the operator must make a decision as to whether the next block of material should be mined and processed, mined and stockpiled, mined (to expose ore) and sent to the waste dump, or not mined at all. The grade of the block is used to make this decision.

在采矿过程中，工人必须做出决定，是否对下一个区域并进行采矿并选矿，对下一个区域进行开采并堆存，或开采（暴露下部矿体）并排送至废石场，或完全不进行开采。

[14] For any block to be deliberately mined, it must pay for the costs of mining, processing, and marketing. The grade of material that can pay for this but for no stripping is the breakeven mining cutoff grade.

任何需要开采的块段，其开采价值必须抵消其采矿成本，选矿成本，市场营销成本。能够达到这个价值的矿石品位就叫盈亏平衡采矿边界品位。

[15] A second cutoff grade can be used for blocks that are below the mining cutoff grade and would not be mined for their own value. These blocks may be mined as waste by deeper ore blocks. The cost of mining these blocks is paid for by the deeper ore.

第二边界品位是低于采矿边界品位，根据其自身的价值就不应该被开采。由于这些块段下部还有高品位矿石需要开采，因此需要剥离掉。剥离这些块段的成本由其下部矿石开采的利润补偿。

[16] The final destination of these blocks is then only influenced by costs for the blocks once

they have been mined. The blocks can be processed at this point if they can pay for just the processing and marketing costs. Because the revenue for the block does not need to cover the mining cost, the milling cutoff grade is lower than the mining cutoff grade.

这样的块段爆破之后是运往选厂还是废石场，要根据其开采成本而定。如果这些块段的品位足够高，只要能补偿其选矿和市场营销的成本，就送往选矿厂。因为这样的块段的价值不需要支付采矿成本，因此，其选矿的边界品位低于采矿边界品位。

Unit 17 Ultimate Pit Definition

最终开采境界的确定

1. Introduction

As the first step for long or short-range planning, the limits of the open pit must be set. The limits define the amount of ore minable, the metal content, and the associated amount of waste to be moved during the life of the operation.

The size, geometry, and location of the ultimate pit are import in planning tailings areas, waste dumps, access roads, concentrating plants, and all other surface facilities. Knowledge gained from designing the ultimate pit also aids in guiding future exploration work.

The material contained in the pit will meet two objectives. A block will not be mined unless it can pay all costs for its mining, processing, and marketing and for stripping the waste above the block. For conservation of resources, any block meeting the first objective will be included in the pit. The result of these objectives is the design that will maximize the total profit of the pit based on the physical and economic parameters used. As these parameters change in the future, the pit design may also change.

2. Manual design

The usual method of manual design starts with the three types of vertical sections shown in Fig. 17.1.

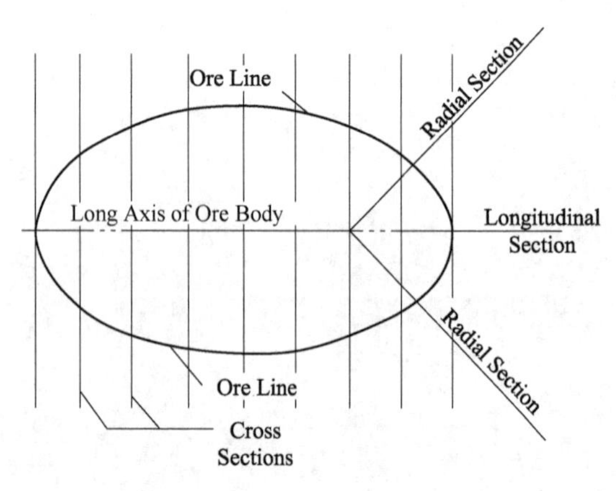

Fig. 17.1 Types of vertical sections used for a manual pit design

Cross sections spaced at regular intervals parallel to each other and normal to the long axis of the ore body. These will provide most of the pit definition and may number from 10 to perhaps 30, depending on the size and shape of the deposit and on the information available. Longitudinal section along the long axis of the ore body to help define the pit limit at the ends of the ore body. Radical sections help define the pit limits at the end of the ore body.

Each section should show ore grades, surface topography, geology (if needed to set the pit limits), and structural controls (if needed to set the pit limits), and any other information that will limit the pit (e. g. ownership boundaries).

The stripping ratio is used to set the pit limits on each section. The pit limits are placed on each section independently using the proper pit slope angle. The pit limits are placed on the section at a point where the grade of ore can pay for mining the waste above it. When the line for the pit limit has been drawn on the section, the grade of the ore along the line is calculated and the lengths of the ore and waste are measured. The ratio of the waste and ore is calculated and compared to the breakeven stripping ratio for the grade of ore along the pit limit. If the calculated stripping ratio is less than the allowable stripping ratio, the pit limit is expanded. If the calculated stripping ratio is greater, the pit limit is contracted. This process continues on the section until the pit limit is set at a point where the calculated and breakeven stripping ratios are equal.

In Fig. 17. 2, the grade on the right side of the pit was estimated to be 0. 6% Cu. At a price of \$2. 25 per kg of copper, the breakeven stripping ratio is 1. 3 : 1.

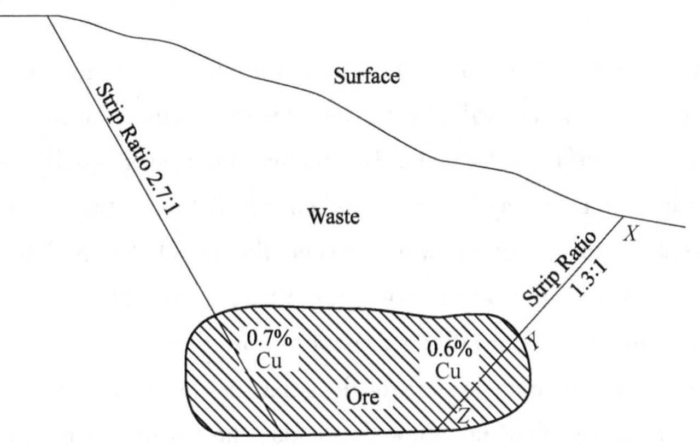

Fig. 17. 2 Pit limits shown on section

The line for the pit limit was found using the required pit slope and located at the point that gave a waste: ore ratio of 1. 3 : 1. At the limit

$$\frac{\text{Length of waste } (XY)}{\text{Length of ore } (YZ)} = \frac{1.3}{1}$$

On the left side of the section, the pit limit for the 0. 7% Cu grade was similarly determined using a breakeven stripping ratio of 2. 7 : 1.

If the grade of ore changed as the pit limit line was moved, the breakeven stripping ratio to use would also change. The pit limits are established on the longitudinal section in the same manner with the same stripping ratio curves.

The pit limits for the radial section are handled with a different stripping ratio curve, however. As shown in Fig. 17.3, the cross sections and the longitudinal section represent a slice along the pit wall with the base length as the surface intercept. The radial section represents a narrow portion of the pit at the base and much wider portion at the surface intercept.

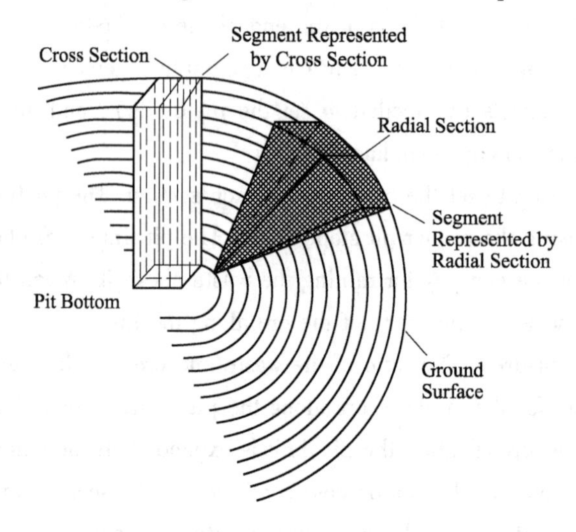

Fig. 17.3 Segments influenced by vertical sections

The next step in the manual design is to transfer the pit limits from each section to a single plan map of the deposit.

The elevation and location of the pit bottom and the surface intercepts from each section are transferred. The resultant plan map will show a very irregular pattern of the elevation and outline of the pit bottom and of the surface intercepts. The bottom must be manually smoothed to conform to the section information. Starting with the smoothed pit bottom, the engineer will develop the outline for each bench at the point midway between the bench toe and the crest. The engineer manually expands the pit from the bottom with the following criteria:

The breakeven stripping ratios for adjacent sections may need to be averaged. The allowable pit slopes must be obeyed. If the road system is designed at the same time, the inter-ramp angle used. If the preliminary design does not show the roads, the outline for the bench midpoints will be based on the flatter overall pit slope that allows for roads.

Possible unstable patterns in the pit should be avoided. Simple geometric patterns on each bench make the designing easier. When the pit plan has been developed, the results should be reviewed to determine if the breakeven stripping ratios have been satisfied.

Vocabulary

cumbersome [ˈkʌmbəsəm] *adj.* 麻烦的；笨重的；难处理的

strike [straɪk] *vt.* 打
n. 走向

cross section	横剖面
longitudinal section	纵剖面
long axis of the ore body	矿体走向
ends of the ore body	矿体端部
radical section	径向剖面
breakeven stripping ratio	经济合理剥采比
surface intercept	露天坑地表境界线
floating cone method	浮锥法
open pit optimization	露天坑优化
constant grade	品位一致

NOTES

[1] As the first step for long or short-range planning, the limits of the open pit must be set. The limits define the amount of ore minable, the metal content, and the associated amount of waste to be moved during the life of the operation.

确定露天开采境界是露天开采长期计划或短期计划编制的基础。露天境界大小决定了坑内开采期间的可采矿石量，金属量和剥离岩石量。

[2] The size, geometry, and location of the ultimate pit are import in planning tailings areas, waste dumps, access roads, concentrating plants, and all other surface facilities. Knowledge gained from designing the ultimate pit also aids in guiding future exploration work.

最终开采境界的大小，几何形态，位置是决定尾矿库，废石场，开拓坑线，选矿厂和其他地表设施的基础。设计的最终境界的资料信息对未来的开采作业起到辅助作用。

[3] The material contained in the pit will meet two objectives. A block will not be mined unless it can pay all costs for its mining, processing, and marketing and for stripping the waste above the block. For conservation of resources, any block meeting the first objective will be included in the pit. The result of these objectives is the design that will maximize the total profit of the pit based on the physical and economic parameters used. As these parameters change in the future, the pit design may also change.

露天坑内的矿岩要满足两个方面要求。如果一个块段的价值不足以抵消其采矿、选矿、市场营销和剥离其上部废石的成本，那么这一块段就不进行采矿作业。从保护资源的角度出发，如果一个块段的价值能抵消其采矿成本，那么这个块段就可以纳入最终境界内。满足这些要求的块段就是最终露天境界的设计。根据矿床自身的开采条件和市场经济条件，设计的露天坑总盈利要达到最大。当未来的市场经济条件发生改变，所设计的露天坑也同样发生改变。

[4] Cross sections spaced at regular intervals parallel to each other and normal to the long axis

of the ore body. These will provide most of the pit definition and may number from 10 to perhaps 30, depending on the size and shape of the deposit and on the information available. Longitudinal section along the long axis of the ore body to help define the pit limit at the ends of the ore body. Radical sections help define the pit limits at the end of the ore body.

矿体的横剖面之间相互平行，各横剖面的间距相等，横剖面与纵剖面垂直。这些横剖面基本确定了露天坑的形状。根据矿床的大小，形态以及相应的其他资料，横剖面可以逐个编号，如10至30。沿矿体走向的纵剖面用于确定最终开采境界在矿体端部的境界。径向剖面用于确定露天坑端部的开采境界。

[5] Each section should show ore grades, surface topography, geology (if needed to set the pit limits), and structural controls (if needed to set the pit limits), and any other information that will limit the pit (e. g. ownership boundaries).

每一个剖面都需要标明矿石品位，地表地形，地质条件（用于确定最终境界），结构控制（用于确定最终境界），和其他可以用于确定最终境界的相关资料（如，领地边界）。

[6] The stripping ratio is used to set the pit limits on each section. The pit limits are placed on each section independently using the proper pit slope angle.

每个剖面的最终开采境界都要用到剥采比。每一个剖面都要根据其稳定的边坡角来单独确定其最终开采境界。

[7] The pit limits are placed on the section at a point where the grade of ore can pay for mining the waste above it.

每个剖面的最终境界都要满足矿石品位具有的价值能抵消其剥离相应废石的成本。

[8] When the line for the pit limit has been drawn on the section, the grade of the ore along the line is calculated and the lengths of the ore and waste are measured.

在最终境界的剖面线画好之后，就要确定该剖面线上矿石的品位，剖面线上矿石的长度和废石的长度。

[9] The ratio of the waste and ore is calculated and compared to the breakeven stripping ratio for the grade of ore along the pit limit.

计算出废石与矿石之比，这个剥采比要与该开采境界线上的矿石品位所具有的盈亏平衡剥采比进行比较。

[10] If the calculated stripping ratio is less than the allowable stripping ratio, the pit limit is expanded. If the calculated stripping ratio is greater, the pit limit is contracted.

如果计算出来的剥采比小于盈亏平衡剥采比，则可以扩大露天境界范围。如果计算出来的剥采比大于盈亏平衡剥采比，则露天境界范围要缩小。

[11] This process continues on the section until the pit limit is set at a point where the calculated and breakeven stripping ratios are equal.

剖面上的计算剥采比与盈亏平衡剥采的比较要持续下去，直至计算出来的剥采比与盈亏平衡剥采比相等。

Unit 18 Open Pit Optimization

露天坑境界优化

1. The management of pit optimization

The first thing to realize is that any feasible pit outline has a dollar value which can, in theory, be calculated. To calculate the dollar value we must decide on a mining sequence and then conceptually mine out the pit, progressively accumulating the revenues and costs as we go.

The second thing to realize is that in doing this calculation we have, in effect, allocated a value to every cubic meter or to every block of rock. Current computer optimization techniques attempt to find the feasible pit outline which has the maximum total dollar value.

2. A simple example

Let us assume that we have a flat topography and a vertical rectangular ore body of constant grade as is shown in Fig. 18. 1. Let us further assume that the ore body is sufficiently long in strike for end effects to be ignored.

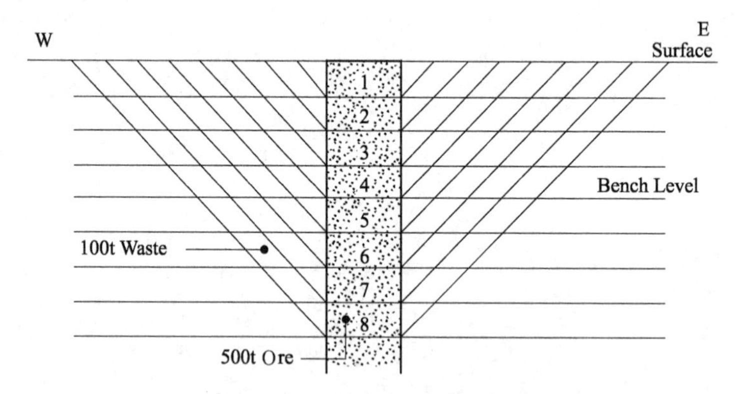

Fig. 18. 1 Simplified ore body

In this simplified case there are eight possible pit outlines that we can consider, and the tonnages for these outlines are given in Table 18. 1.

Table 18. 1 Tonnages for possible pit outlines

pit	ore	waste	total
1	500	100	600
2	1000	400	1400

Continued Table 18. 1

pit	ore	waste	total
3	1500	900	2400
4	2000	1600	3600
5	2500	2500	5000
6	3000	3600	6600
7	3500	4900	8400
8	4000	6400	10400

If we assume that ore is worth \$2 per ton after all mining and processing costs have been paid, and that waste costs \$1 dollar per ton to remove, then we obtain the values shown in Table 18. 2 for the possible pit outlines.

Table 18. 2 Values of the pits if ore is worth \$2 per tonne and if waste costs \$1 per tonne

pit	value
1	900
2	1600
3	2100
4	2400
5	2500
6	2400
7	2100
8	1600

When plotted against pit tonnage, these values produce the graph in Fig. 18. 2. with these very simple assumptions the outline with the highest value is the number five.

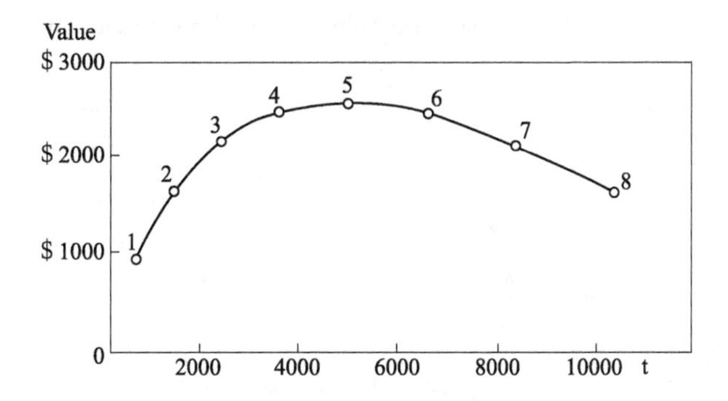

Fig. 18. 2 Relationship between pit tonnage and value

3. The effects of scheduling on the optimal outline

When we schedule a pit, we plan the sequence in which various parts of it will be mined and the time interval in which each is to be mined. This affects the value of the mine because it determines when various items of revenue and expenditure will occur. This is important because the dollar we have today is more valuable to us than the dollar that we are going to receive or spend in a year's time.

In what we will call "worst case" mining, each bench is mined completely before the next bench is started. Waste at the top of the outer shells is mined early.

In what we will call "best case" mining, each shell is mined in turn and thus related ore and waste is mined in approximately the same time period. In this case, the optimal pit is usually close to the one obtained by simple optimization. Unfortunately if we try to mine each shell separately, mining costs usually increase and cancel out some of the gains.

In Fig 18. 3, there are three small ore bodies and their corresponding waste volumes, with their values and costs shown. A floating cone program will examine A and will find that the corresponding cone has a total value of $(40-20-30)=-10$, and so is not worth mining. It will then examine B, will find a cone of value $(200-80-30)=+90$ and will convert it to air, leaving the values shown in Fig. 18. 4.

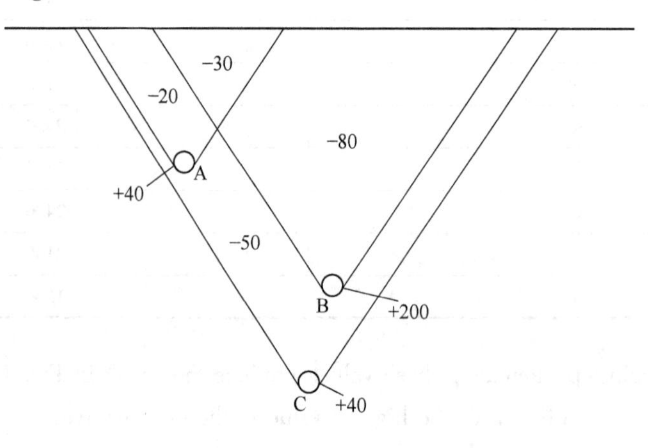

Fig. 18. 3 Ore and waste values before floating cone run

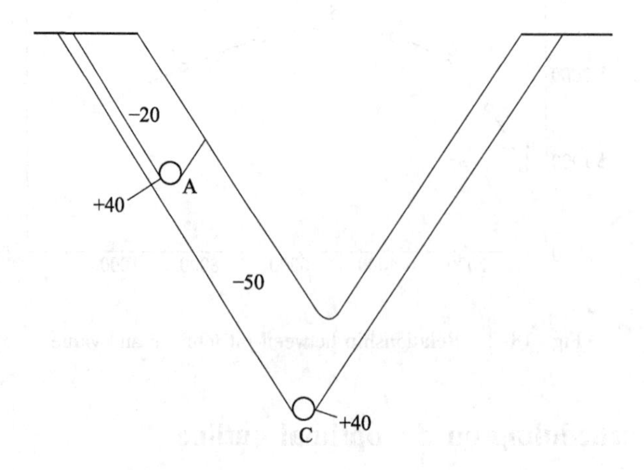

Fig. 18. 4 Ore and waste values after the removal of ore body B and its corresponding waste

The cone for C has a total value of (40−50+40−20) = +10, so that the program mines it. This should not happen, because some of the value of ore body A is being used to help pay for the mining of waste (− 50 region) which is below it. The true optimal pit in this case includes A, B, but not C.

4. Optimum production scheduling

The objective of production scheduling is to maximize the net present value and return on investment that can be derived from the extraction, concentration, and sale of some commodity from an ore deposit. The method and sequence of extraction and the cutoff grade and production strategy will be affected by the following primary factors:

(1) Location and distribution of the ore in respect to topography and elevation;

(2) Mineral types, physical characteristics, and grade/tonnage distribution;

(3) Direct operating expenses associated with mining, processing and converting the commodity into a salable form;

(4) Initial and replacement capital costs needed to commence and maintain the operation;

(5) Indirect costs such as taxes and royalties;

(6) Commodity recovery factors and value;

(7) Market and capital constraints;

(8) Political and environmental considerations.

The procedure used to establish the optimal mining schedule can be divided into three stages. The first defines the extraction order or mining sequence, the second defines a cutoff grade strategy that varies through time and will be optimal for a given set of production parameters, and the third defines which combination of production rates of mine, mill, and refinery will be optimal, within the limits placed by logistical, financial, marketing, and other constraints.

Vocabulary

sequence [ˈsikwəns]	*n.* 顺序 *vt.* 把……按顺序排好
royalties [ˈrɔɪəltiz]	*n.* 版税；矿区土地使用费
processing cost	选矿成本
optimum production scheduling	采掘进度计划优化
pit outline	露天境界
mining sequence	开采顺序
flat topography	地表平坦
floating cone program	浮锥法

NOTES

[1] The first thing to realize is that any feasible pit outline has a dollar value which can, in theory, be calculated. To calculate the dollar value we must decide on a mining sequence and then conceptually mine out the pit, progressively accumulating the revenues and costs as we go.

首先从理论上讲，任何一个露天坑都可以计算出其相应的经济价值。要计算露天坑的开采价值，我们必须确定采矿顺序，然后再去开采各个块段，并累积每个开采块段的经济价值及开采成本。

[2] The second thing to realize is that in doing this calculation we have, in effect, allocated a value to every cubic meter or to every block of rock. Current computer optimization techniques attempt to find the feasible pit outline which has the maximum total dollar value.

其次我们还要知道在计算这些开采块段的利润成本时，我们要给每一个块段的矿岩赋值。目前计算机优化技术可以计算出最终的露天开采境界，使计算出来的露天开采境界的经济价值最大化。

[3] Let us assume that we have a flat topography and a vertical rectangular ore body of constant grade as is shown in Fig. 18. 1. Let us further assume that the ore body is sufficiently long in strike for end effects to be ignored.

假设露天矿地表平坦，矿体的品位均匀，在垂直剖面上成矩形，如图 18. 1 所示。并假设矿体沿走向很长，因此可以忽略端部效应。

[4] If we assume that ore is worth $2 per ton after all mining and processing costs have been paid, and that waste costs $1 dollar per ton to remove, then we obtain the values shown in Table 18. 2 for the possible pit outlines.

我们假定矿石去除采矿和选矿成本之后的价格为 2 美元/t，废石剥离成本为 1 美元/t，则我们可以得出露天坑内的矿床的开采价值，如表 18. 2 所示。

[5] When plotted against pit tonnage, these values produce the graph in Fig. 18. 2. With these very simple assumptions the outline with the highest value is the number five.

根据露天坑内矿岩数量绘制出其价值，如图 18. 2 所示。根据这些简单的假设条件，则价值最大的露天坑为 5 号。

[6] When we schedule a pit, we plan the sequence in which various parts of it will be mined and the time interval in which each is to be mined. This affects the value of the mine because it determines when various items of revenue and expenditure will occur. This is important because the dollar we have today is more valuable to us than the dollar that we are going to receive or spend in a year's time.

当对露天坑的采剥作业进行计划时，我们要确定坑内各个块段矿岩的采剥作业顺序以及各块段之间的采剥作业时间间隔。采剥作业顺序会影响矿山的经济效益，因为采剥作业顺序决定了各个块段的收入和支出的时间。每个块段采剥的时间不同，其价值也不一样，因为今天能得到某一块的矿产品价值要大于后期得到这一块矿产品的价值。

[7] In what we will call "worst case" mining, each bench is mined completely before the next bench is started. Waste at the top of the outer shells is mined early.

每一个台阶采到最终境界之后才进行下一个台阶的开采，这样的开采顺序最差，这种情况下，露天境界最上部的废石会过早剥离。

[8] In what we will call "best case" mining, each shell is mined in turn and thus related ore and waste is mined in approximately the same time period. In this case, the optimal pit is usually close to the one obtained by simple optimization. Unfortunately if we try to mine each shell separately, mining costs usually increase and cancel out some of the gains.

各个台阶轮流采剥作业，在相同的时间段内，采出的矿石量与剥离的废石量保持一定的比例，这种采剥顺序最好。在这种采剥顺序下，最优开采境界非常接近前面介绍的简单实例的优化开采。如果我们一个台阶一个台阶去采剥作业，其采矿成本将会增加，从而使利润变少。

Unit 19 Materials Handling and Waste Disposal

矿岩运输工艺与排岩工艺

1. Introduction

This section will outline design consideration for the development of a hard rock mine in-pit crushing and conveying system. The obvious trend in the nonferrous metals mining industry is toward the mining of lower grade ores at increasing tonnage rates. With the progressive development of larger and more efficient milling equipment and alternative processing techniques the definition of ore, low grade, and waste varies at each operation.

In the past, the primary crushing, fine crushing, and mill complex tended to be located in relative close proximity and the majority of the horizontal and vertical travel distances from the ore source to the crushing station was handled by truck haulage.

Due to rising fuel and maintenance costs, economic conditions have forced the pit designer to minimize the distance that the trucks have to transport the ore, and to bring the primary crusher closer to the source and thus utilize conveyors to perform a much larger proportion of the ore transport requirements.

Data generated from actual installations have shown that properly designed crushing and conveying system compare to truck haulage systems as follows:

(1) Significantly lower operating and maintenance costs.

(2) Higher initial capital costs, but with lower present value costs when compared to the life of the operation.

(3) Improved foul weather operating conditions.

(4) Can provide comparable operating flexibility in certain circumstances.

Studies are currently underway to determine what can be done to improve the technology of in-pit belt conveying. Consideration is being given to the loop belt concept. Many high angle conveying concepts have been studied and the sandwick belt conveyor seems the most promising.

2. Waste disposal

A waste dump is an area in which a surface mining operation can dispose of low grade and / or barren material that has to be removed from the pit to expose higher grade material.

The first step in designing a dump is the selection of a site or sites that will be suitable to handle the volume of waste rock to be removed during the mine's life. Site selection will depend on a number of factors, the most important of which are: (1) Pit location and size through time; (2) Topography; (3) Waste rock volumes by time and source; (4) Property boundaries; (5) Existing drainage routes; (6) Reclamation requirements; (7) Foundation conditions; (8) Material handling equipment. Fig. 19. 1 shows a waste dump.

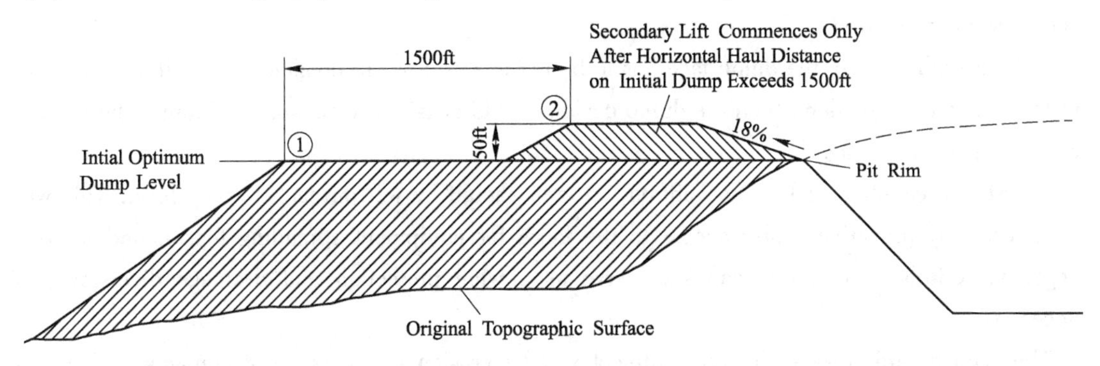

Fig. 19. 1 Equivalent truck haulage distance

Note: Locations ① and ② Would have Equivalent haul costs

3. Stability of mine waste dumps

The overall stability of mine waste dumps depends on a number of factors such as: (1) Topography of the dump site; (2) Method of construction; (3) Geo-technical parameters of mine waste; (4) Geo-technical parameters of the foundation materials; (5) External forces acting on the dump; (6) Rate of advance of the dump face.

All of these factors combine in various ways during the life of a mine waste dump to aid in the stability of the dump or to contribute to its instability.

4. Mine reclamation

The purpose of reclamation is to upgrade the physical character of all or part of a mining area after the mineral values have been removed and, thereafter, to protect the surrounding environment from contamination.

In surface mining operations, the three largest areas that are reclaimed are the mine extraction, the mine waste dump and the mill tailings areas.

If the commodity extracted is a bedded deposit of large extent and of relatively show depth such as in coal mines, the backfilling of worked-out areas is a common method of waste disposal and reclamation. Waste material removed from the initial box cut or pit either be stockpiled and later transported to fill the final excavation or the stockpile could be reclaimed and not be moved and the last pit left with little reclamation effort applied.

In most surface operations in commodities other than coal, the amount of backfilling is restricted of totally impractical. Therefore, most of the reclamation effort is directed toward the waste disposal area.

In most surface mining operations, the waste material removed from the pit is deposited on an adjacent area. The area required for waste disposal is usually equal to or greater than the pit area because the disturbed waste matter has a greater volume than in-situ, a lower slope angle than the pit walls, and rarely can the material be stacked as high as the pit is deep.

In designing waste dumps, particular consideration has to be given to reclamation needs if the cost is to be minimized.

If the overall slope of the dump face has to be reduced to prevent erosion and to allow placement of top soils and vegetation, then the dump design should consider terracing to minimize the amount of material re-handling.

In order to facilitate reclamation efforts, a berm should be left on each terrace level. This will lower costs by providing easier access to the faces for equipment spreading topsoil and for re-vegetation efforts. The berms can also serve as erosion protection and drainage diversions, if necessary.

The main hazards to a reclamation project will be erosion and leakage of contaminated waters that will hamper re-vegetation or be hazardous to life. Both of these problems usually can be corrected through proper drainage control and treatment. Drainage channels will need to be rock-lined if the channels are to remain in the same location without excessive bank erosion.

The depth of topsoil spread on the waste disposal areas will depend to a large extent on the amount that is available. A desirable thickness would be 0. 6m or greater and to a large extent, will depend on reclamation regulations pertaining to the area. In no case should it exceed the current depth of soil cover.

Reclamation of large, deep open pits is very limited if the pits cannot be backfilled. The typical reclamation efforts in these cases would be to allow the pit to fill water. This area then could be used for recreational purposes such as fishing or as a water reservoir. The pit crest can blasted and dozed into the excavation to create a more gentle slope along the edge of the water. This will facilitate water access and will provide a more suitable area for vegetation re-growth and waterfowl habitat. Care has to be exercised using this technique if the chance of bank failure is increased.

Pit excavations can also be used for disposal of other surplus materials, but strict controls have to be applied to avoid ground-water contamination and gaseous emissions.

In cases where the pit is a shallow extraction of large lateral extent above the water table and not backfilled due to a lack of material, the pit can be reclaimed through spreading topsoil and planting suitable vegetation on the pit floor.

5. Ground water and contamination

How the waste material is to be deposited would influence the potential release of contamination from the dump. Of particularly importance from the water pollution standpoint is the permeability of the dump. A waste dump created by end dumping normally tends to be loose and more porous, exhibiting high permeability. On the other hand, a dump constructed in layers exhibits low permeability and has a lower rate of infiltration which, in turn, reduces the risk of contaminating water resources.

From the ground-water impact standpoint, important factors are the permeability characteristics of the soils and bedrock upon which the dump is constructed and the topography of the area. As an example, if the dump is placed on relatively level ground, ample time is provided for infiltration since a perched water level would be created within the dump material. In contrast, a dump placed on steeply sloping ground under similar precipitation conditions would allow less time for infiltration.

The surface water and ground-water measure taken should, of course, comply with applicable regulatory requirements.

6. Surface water

Physical and chemical contamination can be treated through the use of a sedimentation pond. A pond constructed downstream of the dump will retain eroded material transported in the water and collected runoff then is released through an outlet mechanism. For chemical contaminants, it would be essential to determine the amount and type of chemical constituents prior to specifying the necessary treatment.

7. Embankment design

The design of the embankment of the sedimentation ponds should be performed with prudent geotechnical engineering principles and practice. Mine waste, if possible, can be used in the construction of the embankment. The following criteria, as set forth in 30 CFR (Code of Federal Regulations, title 30), should be met in the design of the embankment.

Embankment shall have a minimum static safety factor of 1.5 for the normal pool with steady seepage saturation condition and a seismic safety factor of at least 1.2.

8. Removal of sediment, maintenance and monitoring

Accumulated sediment should be removed from the pond when 60% of the design capacity is utilized. The removed sediment can be hauled to and deposited on the dump. The embankment and appurtenant works should be monitored and maintained so that the structures will continue to function as intended. The settling pond monitoring program should include: (1) Sediment level; (2) Water level and quality; (3) Seepage through embankment and foundation; (4) Sediment accumulation in outlet pipes; (5) Functioning of gates and other mechanical system; (6) Observation of embankment for any signs of deterioration.

9. Ground water contamination and monitoring

Generally metal and nonmetal mine waste will not contribute to the ground water contamination. However, in surface coal mines the material immediately above and below the coal are likely to contain pyritic shale, which may impact on the ground water. In addition to the waste materials, climate, topography, geologic features, and the ground water level below the ground surface can affect the levels of ground water contamination.

A ground water monitoring program should be established to evaluate the impact on ground water. This program should include: (1) Installation of observation wells upstream, downstream and in the vicinity of the waste dump; (2) Water level readings on a regular basis covering wet and dry seasons; (3) Water quality analysis by collecting water samples from the observation wells.

Results of water level fluctuations and water quality dated prior to construction of a waste dump and during the operation should be compared to determine the impact on ground water contamination.

Vocabulary

terracing [ˈterəsɪŋ]	*n.* 梯田，梯级
berm [bəːm]	*n.* 安全平台
dozed [dəuzd]	*v.* 打瞌睡；用推土机清除（挖出，推平）
infiltration [ˌɪnfɪlˈtreʃən]	*n.* 渗透（物）；渗透活动
perched [pɜːtʃt]	*adj.* 停留；被置于高处（或危险处） *v.* 停留，置于（顶上或边上）
runoff [ˈrʌnɒf]	*n.* 径流
embankment [ɪmˈbæŋkmənt]	*n.* 堤防；路基；堤岸；（铁路的）路堤
prudent [ˈprudnt]	*adj.* 谨慎的；小心的；节俭的；深谋远虑的；精明的
appurtenant [əˈpɜːtɪnənt]	*adj.* 附属的；从属的；属于……的 *n.* 附属物，附属的
nonferrous metals	有色金属
the primary crushing	初破碎
fine crushing	细破碎
truck haulage	卡车运输
foul weather	恶劣天气
loop belt	环形运送带
sandwick belt conveyor	夹层运送带
high angle conveying	大角度输送带
barren material	废石
foundation condition	废石场基底条件
rate of advance of the dump face	废石场坡度
drainage diversion	排水改道
rock-lined	衬岩的
water table	潜水面
drainage control	排水控制

pit excavation 露天坑
end dumping 端部倾倒
level ground 水平地面
chemical constituent 化学成分
safety factor 安全系数
seismic safety factor 地震安全系数
appurtenant works 附属工程
pyritic shale 二硫化铁页岩

NOTES

[1] This section will outline design consideration for the development of a hard rock mine in-pit crushing and conveying system. The obvious trend in the nonferrous metals mining industry is toward the mining of lower grade ores at increasing tonnage rates. With the progressive development of larger and more efficient milling equipment and alternative processing techniques the definition of ore, low grade, and waste varies at each operation.

这章介绍硬岩在矿露天坑内的破碎与运输系统设计。非金属矿床开采的发展趋向是开采低品位矿石，且产量越来越大。随着更大、更高效的选矿设备和先进选矿技术的不断发展，矿石、低品位矿石、废石的定义也在不同生产时期相应发生改变。

[2] In the past, the primary crushing, fine crushing, and mill complex tended to be located in relative close proximity and the majority of the horizontal and vertical travel distances from the ore source to the crushing station was handled by truck haulage.

在过去，初破碎、细破碎和综合磨矿的地点多设置在离矿体比较近的地方。从矿体至碎站的大部分水平运输距离和垂直运输作业是由卡车运输完成的。

[3] Due to rising fuel and maintenance costs, economic conditions have forced the pit designer to minimize the distance that the trucks have to transport the ore, and to bring the primary crusher closer to the source and thus utilize conveyors to perform a much larger proportion of the ore transport requirements.

由于燃油和维修费用的提高，露天开采设计者不得不设计出运输距离最短的卡车运输线路，并把破碎站设计在矿体附近，从而可以利用皮带运输大部分矿石。

[4] Data generated from actual installations have shown that properly designed crushing and conveying system compare to truck haulage systems as follows:

(1) Significantly lower operating and maintenance costs.

(2) Higher initial capital costs, but with lower present value costs when compared to the life of the operation.

(3) Improved foul weather operating conditions.

(4) Can provide comparable operating flexibility in certain circumstances.

实际生产数据表明，经过适当设计的破碎站皮带输送系统与卡车运输系统相比具有如下特点：

(1) 作业成本与维修成本大幅降低。

(2) 投资成本增加，但在整个矿山开采期内，其现值成本较低。

(3) 恶劣天气时的作业条件有所改善。

(4) 在有些条件下作业更加灵活。

[5] Studies are currently underway to determine what can be done to improve the technology of in-pit belt conveying. Consideration is being given to the loop belt concept. Many high angle conveying concepts have been studied and the sandwick belt conveyor seems the most promising.

目前正在对相关技术进行研究，以改进坑内皮带运输能力，其中包括考虑采用环形皮带运输的概念来提高运输能力。对许多大角度皮带输送概念也进行过研究，而三明治形状的皮带输送机似乎是最有前途的技术。

[6] A waste dump is an area in which a surface mining operation can dispose of low grade and/or barren material that has to be removed from the pit to expose higher grade material.

废石场是露天开采作业可以处置低品位矿石或废石的地方，只有当这些低品位矿石或废石材料从露天坑内剥离开，才能揭露出高品位矿体。

[7] The overall stability of mine waste dumps is depend on a number of factors such as: (1) Topography of the dump site; (2) Method of construction; (3) Geo-technical parameters of mine waste; (4) Geo-technical parameters of the foundation materials; (5) External forces acting on the dump; (6) Rate of advance of the dump face.

露天矿废石场的整体稳定性取决于以下因素：(1) 废石场的地表地形；(2) 废石场的施工方法；(3) 矿山废石的地质技术参数；(4) 地基材料的地质技术参数；(5) 废石场受到的外力；(6) 废石场工作面推进速度。

Unit 20 Drilling

凿岩

1. Introduction

In virtually all forms of mining, rock is broken through drilling and blasting. Except in dimension stone quarrying, drilling and blasting are required in most surface mining. Only the weakest rock, if loosely consolidated or weathered, can be broken without explosives, using mechanical excavators (ripper, wheel excavators, shovels etc.) or occasionally a more novel device, such as a hydraulic jet.

In the mining cycle, drilling performed for the placement of explosives is termed production drilling. Drilling is also used in surface mining for purposes other than providing blast-holes.

There are minor applications of rock penetration in surface mining other than drilling. In quarrying, dimension stone is freed by cutting, channeling, or sawing. Usually mechanical means or sometimes a thermal jet is employed to produce a cut, outlining the desired size and shape of stone block.

2. Classification of methods

A classification of drilling methods can be made on several bases. These include size of hole, method of mounting and type of power. The scheme that seems the most logical to employ is based on the form of rock attack or mode of energy application leading to penetration.

2.1 Mechanical attack

The application of mechanical energy to rock can be performed basically in only one of two ways: by percussive or rotary action. Combining the two results in hybrid methods termed roller-bit rotary and rotary-percussion drilling.

The mechanical category, of course, encompasses by far the majority (probably 98%) of rock penetration applications today. In surface mining, roller-bit rotaries and large percussion drills are the machines in widest current use, with rotaries heavily favored.

2.2 Thermal attack

The only thermal method having practical application today is flame attack with the jet pierce. It penetrates the rock by spalling, an action associated with hard rocks of high free-silica content. Because of its ready capability of forming various shapes of openings, oxygen or air jet burners are used not only to produce blast holes but to chamber them as well and to cut dimension stone. Jet piercing of blast holes, however, has decreased in popularity in recent years as mechanical drills have improved in versatility and penetrability.

2. 3 Fluid attack

While disintegration of rock by fluid injection is an attractive concept, the end result is more likely fragmentation than penetration. To produce a directed hole with pressurized fluid from an external source, jet action or erosion appears to be more feasible, but commercial application to date is limited.

Hydraulic monitors have been used for over a century to mine placer deposits and to strip frozen overburden, and more recently, high-pressure hydraulic jets have been applied successfully to the mining of coal, and other consolidated materials of relatively low strength.

Hydraulic and mechanical attack mechanisms assist and complement one another. For large holes, the hydraulic jet alone may be competitive with drilling.

2. 4 Sonic attack

Sometimes referred to as vibratory drilling, this method as presently conceived is a form of ultra-high-frequency percussion. Attractive but not presently commercial, actuation of sonic devices by hydraulic, electric, or pneumatic means is possible.

2. 5 Chemical attack

Chemical reaction, because of the time element, may be more attractive as an accessory rather than a primary means of penetration. The use of explosives is a distinct possibility, however, and several alternative systems are under investigation. Additives to the drilling fluid, termed softeners, have shown some improvement in penetration rate in conventional drilling.

2. 6 Other methods of attack

While some attempts to employ other forms of energy (electrical, light, or nuclear) have been made in experimental or hypothetical category at present.

3. Percussion drills

Percussion drills generally plays a minor role as compared with rotary machines in surface mining operations. Their application is limited to production drilling for small mines, second drilling, development work and wall control blasting.

There are two main types of drill mounting. The smaller machines utilize drifter-type drills placed on self-propelled mountings designed to tow the required air compressor. Typical hole sizes are in the 63 to 150mm range (Fig. 20. 1).

The larger machines are crawler-mounted and self-contained. Drill towers permit single pass drilling from 7. 6 to 15. 2m with hole sizes in the range of 120 to 229mm in diameter. These lager

Fig. 20. 1 Percussion drill

machines are almost exclusively operated using down-the-hole hammers.

For many years these machines were exclusively operated using pneumatic hammers. Recently hydraulic machines have been used in the smaller size range. The higher capital cost of these hydraulic drills is offset by lower operating costs and increased productivity compared with pneumatic machines. Another aspect that is becoming increasingly more important is the reduced noise produced by the hydraulic drills.

4. Rotary drills

In rotary drilling, the drill bit attacks the rock with energy supplied to it by a rotating drill stem. The drill stem is rotated while a thrust is applied to it by a pull-down mechanism using up to 65% of the weight of the machine, forcing the bit into the rock. The drill bit breaks and removes the rock by either a ploughing-scraping action in soft rock, or a crushing-chipping action in hard rock, or by a combination of the two.

Compressed air is supplied to the bit via the drill stem. The air both cools the bit and provides a medium for flushing the cuttings from the hole. Water may be used in addition to the compressed air to suppress the effects of dust, however, this is normally found to have a detrimental effects on bit wear. Blast hole sizes produced by rotary machines vary in the range of 100 to 445mm diameter with the most common sizes being 200mm, 250mm, 311mm and 381mm. These drills usually operate in the vertical position (Fig. 20. 2), although many types can drill up to 25 or 300 off the vertical. Drills are manufactured that can drill horizontal holes used in overburden stripping where hard bands of material are located low in the high-wall face.

One of the most important factors in drilling is how fast can drill hole be produced while the machine is actually drilling. This factor almost entirely influences productivity and has a strong influence on unit costs of the hole.

The penetration rate can be expressed by the following empirical equation.

Fig. 20. 2 Rotary drill

$$P = (18.5 - 4.6\lg S_c)\ \frac{W}{\phi} \cdot \frac{N}{5100}$$

where P——penetration rate, m/h;

S_c——uniaxial compressive strength, MPa;

$\frac{W}{\phi}$——weight per meter of bit diameter, t/m;

N——revolution of drill pipe per minute, r/min.

Of the factors in this equation, the rock compressive strength is uncontrollable for a given mine whereas the rotary speed and pull-down can be varied by the drill operator.

The rotary drive motor turns the drill tool string thus turning the drill bit at the bottom of the hole. As the rotary speed increases, so does the number of contacts and the penetration rate. The limit to rotary speed is hot bearings in the bit or stripping of the heel row compacts. Current rotary speeds ranges from 60 to 90r/min for hard materials with greater speeds for softer rocks.

The limitation on penetration rate at many mining properties is the rotary horsepower available. The horsepower requirement can be estimated using the empirical equation:

$$p = \frac{1}{4} K \cdot N \cdot D^{2.5} \cdot W^{1.5}$$

where p——the power of the drive motor, kW;

D——bit diameter, cm;

N——revolution of drill pipe per minute, r/min;

W——weight on the bit, t;

K——constant that varies with rock type. As material strength decreases, the value of K increases. K is about 14×10^{-5} for soft rock down to 4×10^{-5} for high strength materials.

5. Drilling trends

Rotary drills have increased their dominance of blast-hole production in open pit mining. The trend has been to larger sturdier drills, to yield higher mechanical availability and operating performance. This increased availability has been achieved through improvement of crawler track frames, masts, propel chain, pull-down mechanisms, rotary head drives complete with automatic lubrication and greasing, on the new models of machine.

Vocabulary

pierce [pɪəs]	*vt.* 打眼于；穿孔于；打洞于
spall [spɔːl]	*n.* （岩石或矿石的）碎片
	vi. （岩石或矿石）碎裂
	vt. 弄碎（岩石或矿石）
percussion [pəˈkʌʃən]	*n.* 敲击；敲打
hydraulic jet	水射流
placement of explosive	装炸药
rock attack	凿岩
mode of energy	能量形式
mechanical attack	机械凿岩
thermal attack	热力凿岩
fluid attack	水力凿岩
sonic attack	声波凿岩
chemical attack	化学凿岩
hydraulic monitors	水枪
percussion drill	冲击式凿岩机
rotary machine	旋转式凿岩机；牙轮钻机
down-the-hole hammer	潜孔冲击锤

NOTES

[1] In virtually all forms of mining, rock is broken through drilling and blasting. Except in dimension stone quarrying, drilling and blasting are required in most surface mining. Only the weakest rock, if loosely consolidated or weathered, can be broken without explosives, using mechanical excavators (wheel excavators, shovels etc.) or occasionally a more novel device, such as a hydraulic jet.

几乎所有的采矿作业都是通过打眼爆破来破碎岩石的。除采石场外，其他露天开采都要进行凿岩爆破。只有最弱的岩层（如松软破碎或风化岩层）不需要爆破，只需要采用机械挖掘（轮式挖掘机，电铲等），偶尔也采用更新颖的水力喷射设备开采。

[2] In the mining cycle, drilling performed for the placement of explosives is termed production drilling. Drilling is also used in surface mining for purposes other than providing blast-holes.

在采矿过程中，为了装填炸药所进行的穿孔凿岩作业称生产穿孔凿岩。在露天采矿过程中，穿孔除了用于装填炸药之外，还有其他方面的作用。

[3] There are minor applications of rock penetration in surface mining other than drilling. In quarrying, dimension stone is freed by cutting, channeling, or sawing. Usually mechanical means or sometimes a thermal jet is employed to produce a cut, outlining the desired size and shape of stone block.

露天采矿中，多数凿岩孔用于爆破，少数凿岩孔不用于爆破。在采石场是通过切割、穿孔、拉锯的方法获得石材的。通常采用机械方法，有时采用喷热方法来切割岩石，得到所需要的石材形状。

[4] A classification of drilling methods can be made on several bases. These include size of hole, method of mounting and type of power. The scheme that seems the most logical to employ is based on the form of rock attack or mode of energy application leading to penetration.

穿孔方法可以分成几类。包括孔的大小，钻机架设方法和动力类型。最合逻辑的分类方法是凿岩的方式或凿岩的能量形式。

[5] The application of mechanical energy to rock can be performed basically in only one of two ways: by percussive or rotary action. Combining the two results in hybrid methods termed roller-bit rotary and rotary-percussion drilling.

机械能凿岩基本上是采用如下两种凿岩方式之一：冲击式凿岩和旋转式凿岩。把这两种凿岩的方式结合在一起就是牙轮钻机凿岩或旋转冲击式凿岩。

[6] The mechanical category, of course, encompasses by far the majority (probably 98%) of rock penetration applications today. In surface mining, roller-bit rotaries and large percussion drills are the machines in widest current use, with rotaries heavily favored.

当然，目前绝大部分凿岩是机械式凿岩（约占98%）。当今的露天采矿中，牙轮钻机和大型冲击式钻机应用最广，其中企业更喜欢使用牙轮钻。

[7] The only thermal method having practical application today is flame attack with the jet pierce. It penetrates the rock by spalling, an action associated with hard rocks of high free-silica content. Because of its ready capability of forming various shapes of openings, oxygen or air jet burners are used not only to produce blast holes but to chamber them as well and to cut dimension stone. Jet piercing of blast holes, however, has decreased in popularity in recent years as mechanical drills have improved in versatility and penetrability.

目前唯一使用热力冲击破岩的方式是采用喷热式冲击破岩。喷热切削岩石成孔，这种喷热成孔作业在含硅高的岩石中进行。由于喷热冲击破岩容易形成不同形状的空孔，喷氧或喷气枪既可用于钻凿炮孔，也可以在岩石四周切削，获得方形石材。然而，如今喷热凿岩逐渐减少，而机械凿岩的效率提高，应用更广。

[8] While disintegration of rock by fluid injection is an attractive concept, the end result is more likely fragmentation than penetration. To produce a directed hole with pressurized fluid from an external source, jet action or erosion appears to be more feasible, but commercial application to date is limited.

水力喷射凿岩是一种很有富有想象力的方法。水力喷射凿岩更像是破碎岩石而不是穿孔。外部喷射的压力水流朝一个方向喷射是可以形成炮孔的，但是，到目前为止，这方面的商业运作有限。

[9] Hydraulic monitors have been used for over a century to mine placer deposits and to strip frozen overburden, and more recently, high-pressure hydraulic jets have been applied successfully to the mining of coal, and other consolidated materials of relatively low strength.

水力喷射枪由于开采砂积矿床，剥离表土冰层已有一百多年的历史。近年来，高压水流喷射已应用于煤炭开采和其他强度相对较小的固体矿床开采。

[10] Hydraulic and mechanical attack mechanisms assist and complement one another. For large holes, the hydraulic jet alone may be competitive with drilling.

水力凿岩与机械凿岩互补。钻凿大孔时，水力凿岩的成孔效率更高。

[11] Sometimes referred to as vibratory drilling, this method as presently conceived is a form of ultra-high-frequency percussion. Attractive but not presently commercial, of sonic devices by hydraulic, electric, or pneumatic means is possible.

声波凿岩有时也叫振动凿岩。声波凿岩是近期构思的一种超高频率冲击式凿岩。声波凿岩设备可以采用水力，电力或风力作为动力，这种设备的凿岩方式具有想象力，但是尚无实际商业应用。

[12] Chemical reaction, because of the time element, may be more attractive as an accessory rather than a primary means of penetration. The use of explosives is a distinct possibility, however, and several alternative systems are under investigation. Additives to the drilling fluid, termed softeners, have shown some improvement in penetration rate in conventional drilling.

化学凿岩可以作为一种辅助凿岩手段，而非主要凿岩手段。由于化学反应的时间因素，因此，化学凿岩更加具有想象力。利用炸药凿岩明显可行，此外，其他形式的化学凿岩还在调查研究之中。实验表明，在凿岩的流体中加入某种化学物质（即软化剂）可以提高穿孔效率。

[13] While some attempts to employ other forms of energy (electrical, light, or nuclear) have

been made in experimental or hypothetical category at present.

其他形式的凿岩（电力，光力，原子能）已经做了实验性研究或提出了假设性的实验方法。

[14] In rotary drilling, the drill bit attacks the rock with energy supplied to it by a rotating drill stem. The drill stem is rotated while a thrust is applied to it by a pull-down mechanism using up to 65% of the weight of the machine, forcing the bit into the rock. The drill bit breaks and removes the rock by either a ploughing-scraping action in soft rock, or a crushing-chipping action in hard rock, or by a combination of the two.

旋转式凿岩中，钻头依靠旋转钻杆提供的能量冲击岩石。钻杆旋转的同时，下拉机构给钻杆施加65%的机器自重的推力，迫使钻头切进岩石中。钻头通过在软岩中的切削作用或在硬岩中的挤压破碎作用或者切削挤压破碎同时作用对岩石进行破碎并移动岩屑。

Unit 21 Blasting

爆破

1. Introduction

Most rocks require blasting prior to excavation in surface mining operations. Blasting is the most important of the unit operations for many mines because if it is not performed successfully, the viability of the mine frequently becomes jeopardized.

The principal factors that influences blasting results are the properties of the explosives being used, their distribution and initiation sequence in the blast, the overall blast geometry, and the rock structure and other properties.

This section indicates the effects of these different factors and describes suitable blast design techniques. In addition, specialized blasting methods such as throw blasting and wall control are discussed along with some of the detrimental aspects of blasting, vibration, air blast, and noise.

2. Pre-splitting

Fig. 21. 1 illustrates typical pre-split blast layout using 102mm diameter pre-split holes for 250mm diameter production holes.

First this type of blast, pre-split holes would normally be drilled first, ahead of main production holes. The choice can then be made between loading and firing the pre-split line or infilling the main blast. In the latter case, the pre-split line would be fired instantaneously 100 to 150 milliseconds before the main blast. As shown in Fig. 21. 1, the pre-split line is formed ahead of the main blast and allows the gas being driven back from the buffer row through the radial cracks to terminate at the pre-split line.

The pre-split row in Fig. 21. 1 has a spacing of 2m for a 102mm diameter hole and is inclined at 15° to the vertical. The pre-split angle is somewhat dictated by rock structure although a slight angle is preferred regardless of structure for long-term stability as well as for best initial results with large production holes.

The Fig. 21. 1 illustrates the upper bench where two benches will finally run together to form the final face between berms.

Pre-split drill requirements become clear when pre-split holes needed for the next bench are considered.

The drill must be capable of drilling close to the previously produced bench face at an angle of 15° beneath itself so the face can be continued to depth. Currently, this means some form of drifter

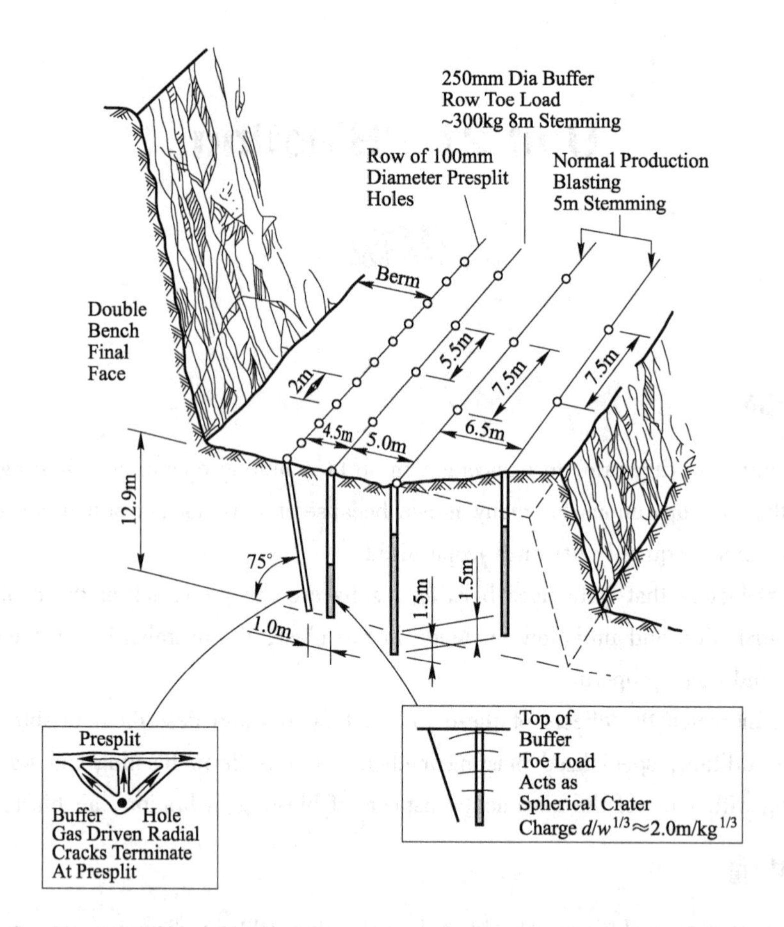

Fig. 21. 1 Presplit blast coupled to a 250mm diameter production blast

drill is required limiting the hole size to 102 to 127mm diameter.

The back row of the main production blast, termed the buffer row, must also be carefully designed with respect to standoff distance from the pre-split row and spacing as well as explosive load.

The inset sketch on the right side of Fig. 21. 1 shows the top portion of the buffer row hole charge acts as a spherical crater charge breaking to the bench surface. Subsequently, main blast holes after the buffer row are designed at regular spacing, burden, and loading for the type of material blasted.

One further point to note from Fig. 21. 1 is the sub-grade, more accurately, lack of sub-grade used on the pre-split and used on the pre-split and buffer row holes. This is to prevent damage to the bench below or to the wall at that point.

To minimize cratering from the top of the pre-split charge, the collar should be such as to avoid cratering. In massive brittle rock, the scaled burial depth of the top eight charge diameters should be 4m or greater. In softer rock it can be as high as 5m.

The buffer row is the row of holes ahead of the pre-split line. When the buffer row is designed, it is essential that the top of the charge is contained to such a degree that it will not crater at the top beyond the pre-split line which is fired first.

The pre-split line is fired either prior to drilling the main blast or 100 to 150 millisec ahead of the production rows.

3. Trim blasting

Trim blasting is design technique to give wall control using large diameter rotary blast holes for both production and final row holes. The idea, then, is to eliminate costly small diameter blast hole work, along with the associated hole loading difficulties.

Fig. 21. 2 schematically presents a crest trim and wall trim blast. The upper blast, called the crest trim blast, takes the upper bench to limit. From Fig. 21. 2 the majority of backbreak occurs at the bench crest mostly from previous subgrade.

The blast has three distinct components, similar to the pre-split blast.

A trim row is used to produce the final wall in a similar manner to the pre-split row. The trim row is decoupled.

If stemming is used, then the decoupling calculation is performed using 40% to 50% voids. Again, a buffer row is designed as the last row of the main blast, with increased stemming to prevent cratering back at the surface through the trim row.

Normally, two other regular rows of holes would be used in front of the buffer row to complete the trim blast. These two rows would be at normal spacing and burden and would be loaded using the standard procedure for the appropriate material type.

All holes would be vertical and would be of production size. For example, in the case of Fig. 21. 2, holes are 250mm diameter.

Spacing generally ranges from 12 to 16 times the hole diameter. These values correspond to a range or rock hardness from hard to moderately soft and are for adequately decoupled charges.

The second bench blast used to complete the two bench face, termed the wall trim blast in Fig. 21. 2, is shown on the lower portion of the figure. The blast is similar to the crest trim blast, except the trim row is 3m from the toe position. This distance is dictated by drilling equipment size. Some backbreak from the wall trim blast makes the final wall almost continuous, with a slight ripple remaining where the two benched join. It should be noted that this trim blasting technique does not prove effective where the rock structure dips at a shallow angle into the pit.

A number of mines are now attempting to achieve minimum overall cost by adjusting their drilling and blasting cost to the optimum. In many cases the optimization includes pre-splitting and throw blasting.

One major reason for pre-splitting, apart from improved safety, occurs at operations with water problems. The pre-split commonly dewaters the blastholes, which enables easy use of dry mix explosives rather than the more expensive slurries. When pre-splitting, a 50m wide pit for a 300m length, the entire pre-split block commonly moves up to 1m towards the mines cut. Main blast holes drilled after the pre-split has been fired are usually dry, even in areas expected to be filled with water after drilling.

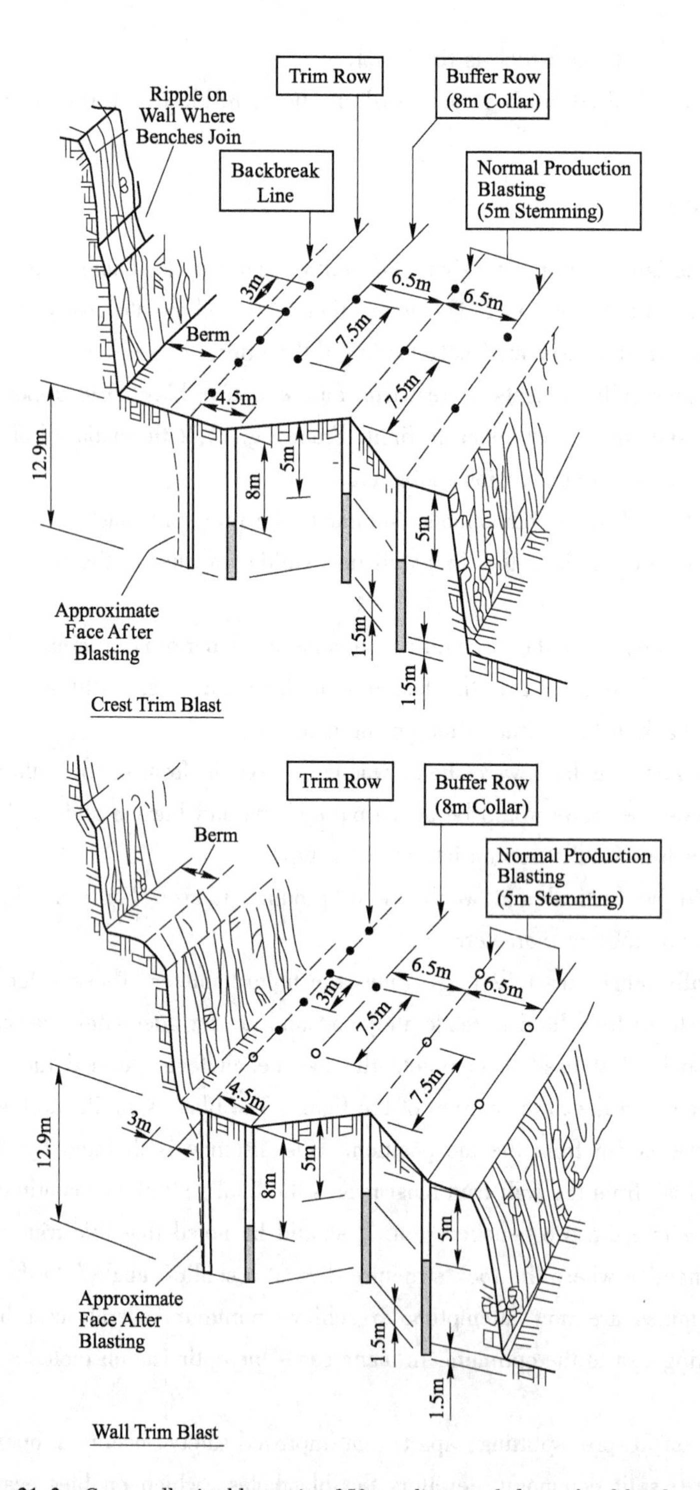

Fig. 21.2 Crest wall trim blast using 250mm diameter holes in hard brittle rock

A pre-split face allows accurate control on the burden for the full length of the front row blast holes, allowing more than 60% of the material to be throw clear of the coal at some operation.

4. Crater blasting

Cratering using spherical charges finds wide application in surface mining operations. For example, crater blasting test results are commonly used to design the correct blast hole collars either for fragmentation or flyrock control.

5. Blast vibration and air blast

The principal factors that affect vibration levels at a given point of interest are the weight of explosive fired, the distance from the blast, the delay period, if any, and the blast geometry.

The scaled distance combines the effects of charge weight (W) on the geometrical dispersion of the vibration at distance (d) in the form of $d/W^{\frac{1}{2}}$.

An empirical equation of the form

$$V_{max} = K\left(\frac{d}{W^{\frac{1}{2}}}\right)^m$$

Relating the peak particle velocity with scaled distance must therefore be developed. K and m are related to the geological characteristics of the rock and delay period.

5. 1 Vibration effects

The problem of vibration from blasting is one of the most pressing facing mine operators in populated areas, both from a potential structural damage claim and from the standpoint of the annoyance effect on people.

5. 2 Blast vibration measurement

In the measurement of vibration, the objective is to detect and record the vibratory motion of the ground. The quantities measured must results in a full description for the vibratory event. This requires three orthogonal components of either particle displacement, velocity, or acceleration.

For typical blast vibrations it is necessary that seismometer should be able to record the following ranges:

Frequency, 1 to 100Hz.

Displacement, 0. 00254 to 12. 7mm.

Velocity, 0. 254 to 254mm/s.

Acceleration, 0. 005 to 2g.

6. Noise and air blast

Whenever a pressure wave travels through the air faster than a sound wave, it produces a shock wave. The airborne pressure wave emanated from an explosion, called air blast, is a shock wave. At first it travels at supersonic speed, but, depending on the magnitude of the energy released by the explosion, it will decay in time to an ordinary sound wave.

Noise and sound is a pressure wave traveling through air at approximately 335m/s.

Air blast acts in a similar manner to noise or sound as it is also a pressure-type wave. However it has a greater speed than of sound and its frequency is less than 20Hz, so consequently it cannot be heard. Its effect is apparent in the rattling of windows, etc.

7. Air blast control

In the pit blasting, noise or air blast can be due to any of the following items:

primacord truck lines;

lack of proper stemming material;

inadequate stemming height;

overdug or overloaded front row holes;

burden near the crest too small due to backbreak;

delay sequence;

atmospheric conditions such as temperature inversions or wind in the direction of concern.

Vocabulary

viability [ˌvaɪəˈbɪlɪtɪ] *n.* 生存力

jeopardize [ˈdʒepəˌdaɪz] *vt.* 使处于危险境地；危害；损害

burden [ˈbɜːdn] *n.* 负荷；抵抗线
vt. 加负荷于；使负担

collar [ˈkɒlə] *n.* 颈；轴环；孔口

crater [ˈkreɪtə] *n.* 火山口；弹坑；爆破漏斗

backbreak [ˈbækbreɪk] *n.* 后冲

ripple [ˈrɪpl] *n.* 涟漪；波纹
vi. 起涟漪；起波纹

seismometer [saɪzˈmɒmɪtə] *n.* 地震计

decay [dɪˈkeɪ] *n.* 衰变
vi. （原子核）衰变；衰变

overdug [ˈəʊvərˈdʌg] *n.* 超深

initiation sequence 起爆顺序

overall blast geometry 炮孔布置

throw blasting 抛掷爆破

buffer row 缓冲排炮孔

radial cracks 径向裂隙

pre-splitting 预裂

spacing 孔间距

sub-grade 超深

double bench final face 并段台阶坡面

toe load	孔底装药量
spherical crater charge	球状药包装药
trim blast	光面爆破
wall trim blast	并段光面爆破
crest trim blast	台阶光面爆破
toe position	坡底线
casting in dragline operations	索斗铲装作业
blast vibration and air blast	爆破振动与空气冲击
delay period	延时时间
blast geometry	爆破地质条件
flyrock control	飞石控制
particle displacement, velocity, or acceleration	质点的位移，速度及加速度
air blast	空气冲击波
pressure-type wave	压力波
rattling of windows	窗户咯咯作响
primacord truck lines	主导爆索
stemming material	堵塞材料
stemming height	堵塞长度

NOTES

[1] Most rocks require blasting prior to excavation in surface mining operations. Blasting is the most important of the unit operations for many mines because if it is not performed successfully, the viability of the mine frequently becomes jeopardized.

多数岩石需要在采装之前进行爆破。爆破是许多矿山最重要的生产环节，如果爆破工作没做好，矿山就难以维持下去。

[2] The principal factors that influences blasting results are the properties of the explosives being used, their distribution and initiation sequence in the blast, the overall blast geometry, and the rock structure and other properties.

影响爆破效果的主要因素是使用的炸药性能，炸药的分配，起爆顺序，炮孔布置，岩石结构及性能。

[3] This section indicates the effects of these different factors and describes suitable blast design techniques. In addition, specialized blasting methods such as throw blasting and wall control are discussed along with some of the detrimental aspects of blasting, vibration, air blast, and noise.

这章讲述这些因素对爆破质量的影响程度，并讨论适用的爆破技术。此外，还要讨论控制爆破（如抛掷爆破，光面爆破）以及爆破振动，空气冲击和爆破噪声的危害。

[4] Fig. 21. 1 illustrates typical pre-split blast layout using 102mm diameter pre-split holes for 250mm diameter production holes.

图 21. 1 是典型的预裂爆破布置设计，该设计采用 102mm 直径的预裂孔和 250mm 的生产炮孔。

[5] First this type of blast, pre-split holes would normally be drilled first, ahead of main production holes. The choice can then be made between loading and firing the pre-split line or infilling the main blast. In the latter case, the pre-split line would be fired instantaneously 100 to 150 milliseconds before the main blast. As shown in Fig. 21. 1, the pre-split line is formed ahead of the main blast and allows the gas being driven back from the buffer row through the radial cracks to terminate at the pre-split line.

首先，这种爆破设计中，预裂孔在生产主爆孔之前进行钻凿。然后再决定是先在预裂孔里装药并起爆预裂孔，再在主爆孔里装药爆破，还是在预裂孔和主爆孔里都装药进行爆破。在后面这种情况下，预裂孔与主爆孔同次分段爆破，且预裂孔要在主爆孔之前 100 至 150ms 起爆。如图 21. 1 所示，预裂缝在主爆孔爆破之前就形成，使得缓冲排炮孔的径向裂隙在预裂缝处终止，并且使得缓冲排炮孔的爆轰气体通过径向裂隙返回。

[6] The pre-split row in Fig. 21. 1 has a spacing of 2m for a 102mm diameter hole and is inclined at 15° to the vertical. The pre-split angle is somewhat dictated by rock structure although a slight angle is preferred regardless of structure for long-term stability as well as for best initial results with large production holes.

图 21. 1 中的预裂孔的孔间距是 2m，炮孔直径为 102mm，与垂直面的倾斜角度为 15°。预裂孔的倾角与岩石结构有关。无论岩石结构如何好，从边坡长期稳定的角度考虑，预裂孔稍微倾斜是有益边坡稳定，同时也利于生产炮孔最初期的布置。

[7] The Fig. 21. 1 illustrates the upper bench where two benches will finally run together to form the final face between berms.

从图 21. 1 还可以看出，上部台阶是两个台阶合并的结果，并在两个安全平台之间形成了台阶坡面。

[8] The drill must be capable of drilling close to the previously produced bench face at an angle of 15° beneath itself so the face can be continued to depth. Currently, this means some form of drifter drill is required limiting the hole size to 102 to 127mm diameter.

预裂孔必须靠近上一个台阶坡面，并与垂直面成 15°，如此就能使预裂面保持连续向下。因此，良好的预裂面需要边坡钻机的钻孔直径控制在 102～127mm。

[9] The back row of the main production blast, termed the buffer row, must also be carefully designed with respect to standoff distance from the pre-split row and spacing as well as explosive load.

主爆孔后面一排炮孔叫缓冲孔。缓冲孔的设计要仔细考虑缓冲孔与预裂孔的距离，缓冲孔之间的孔距以及缓冲孔的装药情况。

[10] The inset sketch on the right side of Fig. 21.1 shows the top portion of the buffer row hole charge acts as a spherical crater charge breaking to the bench surface. Subsequently, main blast holes after the buffer row are designed at regular spacing, burden, and loading for the type of material blasted.

图 21.1 右边的草图中可以看出，缓冲孔装药的上部就像一个爆破漏斗，炸药爆破朝台阶面推进。因此，在爆区中，缓冲孔后面的主爆孔的孔间距，抵抗线和装药量要一致。

[11] One further point to note from Fig. 21.1 is the sub-grade, more accurately, lack of sub-grade used on the pre-split and buffer row holes. This is to prevent damage to the bench below or to the wall at that point.

图 21.1 中还有一点需要说明，即预裂孔和缓冲孔不设超深。这是为了避免破坏下一个台阶或者是避免破坏此处的台阶面。

[12] Trim blasting is design technique to give wall control using large diameter rotary blast holes for both production and final row holes. The idea, then, is to eliminate costly small diameter blast hole work, along with the associated hole loading difficulties.

光面爆破是控制台阶坡面的设计技术，爆破采用牙轮钻钻凿的大孔径炮孔进行爆破，包括生产炮孔和最后一排光面孔。采用光面爆破的原因是不使用钻孔成本高的小钻孔凿岩，同时也避免了小孔装药的困难。

[13] Fig. 21.2 schematically presents a crest trim and wall trim blast. The upper blast, called the crest trim blast, takes the upper bench to limit. From Fig. 21.2 the majority of backbreak occurs at the bench crest mostly from previous subgrade.

图 21.2 系统给出了台阶光面爆破和并段光面爆破的设计。图 21.2 中上部的爆破，即台阶光面爆破，其爆破以上一个台阶为限。从图 21.2 可以看出，大部分后冲形成的坡顶线位置在以前超深孔处。

[14] A trim row is used to produce the final wall in a similar manner to the pre-split row. The trim row is decoupled.

最后，光面孔通常会形成与预裂爆破一样的台阶坡面。光面孔采用不耦合装药。

[15] If stemming is used, then the decoupling calculation is performed using 40% to 50% voids. Again, a buffer row is designed as the last row of the main blast, with increased stemming to prevent cratering back at the surface through the trim row.

如果对光面孔进行堵塞，则不装药的空间为 40%~50%。而且，缓冲孔作为主爆孔的最后一排，其堵塞深度要比正常炮孔的堵塞深度大，目的是避免产生漏斗爆破作用，破坏后面的光面层。

[16] Normally, two other regular rows of holes would be used in front of the buffer row to complete the trim blast. These two rows would be at normal spacing and burden and would be loaded using the standard procedure for the appropriate material type.

通常在缓冲孔前面布置两排正常生产炮孔，方便形成光面层。这两排孔的孔间距与抵抗线以及装药情况与生产炮孔一样。

[17] All holes would be vertical and would be of production size. For example, in the case of Fig. 21.2, holes are 250mm diameter.

所有炮孔都垂直布置，且光面孔、缓冲孔、主爆孔的直径都一致。如图 21.2 所示，所有炮孔直径都是 250mm。

[18] The second bench blast used to complete the two bench face, termed the wall trim blast in Fig. 21.2, is shown on the lower portion of the figure. The blast is similar to the crest trim blast, except the trim row is 3m from the toe position. This distance is dictated by drilling equipment size. Some backbreak from the wall trim blast makes the final wall almost continuous, with a slight ripple remaining where the two benched join. It should be noted that this trim blasting technique does not prove effective where the rock structure dips at a shallow angle into the pit.

如图 21.2 下半部分所示，在第二个台阶爆破之后完成两个台阶并段，称为并段光面爆破。并段光面爆破与台阶光面爆破类似，但是当一个台阶的光面孔离上一台阶坡底线达 3m 之远，就不算并段光面爆破了。两个台阶之间的距离与钻机设备型号大小有关。并段光面爆破的后冲使得并段后的台阶坡面看上去为一个连续的坡面，只在两个台阶坡面结合处留下一个小的搭接不整合段。应该注意的是，这种并段光面爆破技术在岩石结构面倾斜角度小的边坡上实施的效果不好。

[19] To minimize cratering from the top of the pre-split charge, the collar should be such as to avoid cratering. In massive brittle rock, the scaled burial depth of the top eight charge diameters should be 4m or greater. In softer rock it can be as high as 5m.

为了避免预裂爆破孔形成爆破漏斗，预裂孔孔口的填塞长度要足够长。在脆性岩体中，孔口部位的填塞深度为 4m，或者为八倍以上的炮孔直径。在软岩中，填塞长度则长达 5m。

[20] The buffer row is the row of holes ahead of the pre-split line. When the buffer row is designed, it is essential that the top of the charge is contained to such a degree that it will not crater at the top beyond the pre-split line which is fired first.

缓冲排炮孔是指预裂孔之前的一排孔。如果设计缓冲孔，那么缓冲孔堵塞长度应该足够长，使缓冲孔爆破不在顶部形成漏斗，从而不破坏最先起爆的预裂面。

[21] The pre-split line is fired either prior to drilling the main blast or 100 to 150ms ahead of the production rows.

预裂排孔可以在主爆孔钻凿之前起爆，或者在主爆孔爆破之前 100~150ms 起爆。

[22] One major reason for pre-splitting, apart from improved safety, occurs at operations with water problems. The pre-split commonly dewaters the blastholes, which enables easy use of dry mix explosives rather than the more expensive slurries. When pre-splitting, a 50m wide pit for a 300m length, the entire pre-split block commonly moves up to 1m towards the mines cut. Main blast holes drilled after the pre-split has been fired are usually dry, even in areas expected to be filled with water after drilling.

预裂爆破除了能提高边坡稳定，还有防水的功效。预裂缝通常可以排出炮孔水，从而可以使用价格便宜的干混合炸药，而无须使用成本高的浆状炸药。当产生 50m 宽，300m 长的预裂缝之后，整个预裂爆区通常向坑内移动长达 1m 的距离。预裂孔爆破之后，再钻凿主爆孔时，主爆孔通常是干涸的，甚至在过去曾经充满水的区域钻孔也出现干涸的钻孔。

References 参考文献

[1] 蒋国安，吕家立. 采矿工程英语［M］. 北京：中国矿业大学出版社，1998.

[2] 周科平，李杰林. 采矿工程专业英语［M］. 长沙：中南大学出版社，2010.

[3] 杨科. 采矿工程专业英语［M］. 北京：中国科学技术大学出版社，2010.

[4] 卢宏建，姚旭龙. 采矿工程专业英语［M］. 北京：冶金工业出版社，2016.

[5] William A Hustrulid, Richard L Bullock. Underground Mining Methods [M]. Society for Mining, Metallurgy, and Exploration, Inc., 2001.

[6] Howard L Hartman, Jan M Mutmansky. Introductory Mining Engineering [M]. Wiley, 2002.

[7] Bruce A Kennedy. Surface Mining [M]. Society for Mining, Metallurgy, and Exploration, Inc., 1988.

[8] Peter Darling. SME Mining Engineering Handbook (3rd Edition) [M]. Society for Mining, Metallurgy, and Exploration, Inc., 1992.

[9] Evert Hock, Ted Brown. Underground Excavations in Rock [M]. CRC Press, 1980.

[10] Donald G. Newman, Ted G. Eschenbach, Jerome P. Lavelle, Neal A. Lewis. Engineering Economic Analysis [M]. Oxford University Press, USA, 2019.

[11] Richard E. Goodman. Introduction To Rock Mechanics (2nd Edition) [M]. John Weily & Sons, Inc., 1989.

[12] Gokhale, Bhalchandra V. Rotary Drilling and Blasting in Large Surface Mines [M]. CRC Press, 2018.

[13] Albert Williams Daw, Zacharias Williams Daw. The Blasting of Rock in Mines, Quarries, Tunnels, Etc [M]. Kessinger Publishing, 2010.

[14] Gupta A, Yan D S. Mineral Processing Design and Operation: An Introduction [M]. Elsevier Science, 2006.

[15] 吴春荣，张锡濂. 英汉地质学缩写词汇［M］. 北京：科学出版社，1978.

[16]《矿业词汇》编审委员会. 简明英汉矿业词汇［M］. 北京：煤炭工业出版社，1997.

[17] 宦秉炼. 地矿英汉词典［M］. 北京：冶金工业出版社，2020.